Military Soft Power

Military Soft Power

Public Diplomacy through Military Educational Exchanges

Carol Atkinson

ROWMAN & LITTLEFIELD
Lanham • Boulder • New York • Toronto • Plymouth, UK

Published by Rowman & Littlefield
4501 Forbes Boulevard, Suite 200, Lanham, Maryland 20706
www.rowman.com

10 Thornbury Road, Plymouth PL6 7PP, United Kingdom

British Library Cataloguing in Publication Information Available

Library of Congress Cataloging-in-Publication Data

Atkinson, Carol L., 1959–
Military soft power : public diplomacy through military educational exchanges / Carol Atkinson.
pages cm
Includes bibliographical references and index.
ISBN 978-1-4422-3128-3 (cloth : alk. paper)—ISBN 978-1-4422-3129-0 (electronic)
1. Military education—United States—Political aspects. 2. Armed forces—Officers—Training of. 3. Students, Foreign—United States. 4. Educational exchanges—United States. 5. United States—Military relations. 6. Democratization—Government policy—United States. I. Title.
U408.3.A85 2014
355.5'50973—dc23 2014006876

™ The paper used in this publication meets the minimum requirements of American National Standard for Information Sciences—Permanence of Paper for Printed Library Materials, ANSI/NISO Z39.48-1992.

Printed in the United States of America

Contents

List of Tables

List of Figures

Acknowledgments

I am very thankful to all of the international exchange officers who generously took the time to complete my survey as well as the many people who helped administer the survey. Without them, the research presented in this book would not have been possible. The officers will continue to remain anonymous, but not the very helpful staff at the war and staff colleges. At the U.S. National Defense University, I thank John Yaeger, John Charlton, Paul Lambert, Mark McGivern, and Adam Jungdahl. At the U.S. Army Command and General Staff College I thank Jim Fain, Maria Clark, and Jeremy Dunn. At Air University I thank Mike Homan, George McCarthy, Jr., Edward Brown, Jerry Gantt, Douglas Peifer, Kathleen Mahoney-Norris, Matthew Schwonek, Mary Hampton, Janet Beilstein, Arden Gale, Steve Hansen, and Joe Lanzetta. At the U.S. Marine Corps University I thank Robert Liston and Jerre Wilson. At the U.S. Army War College I thank Alex Portelli, Anna Waggener, and Marybeth Peterson Ulrich. I am also very fortunate for the friendship of my international colleagues Zé Mendonça, Jaro Svacho, and João Barbosa who have stayed in touch throughout the last fifteen years enriching my own life and adding their unique perspectives to it.

I would like to extend my deep gratitude to Giacomo Chiozza, Jim Fain, and Kathleen-Mahoney Norris. I am very grateful to Giacomo and Jim who have provided enthusiastic support, insightful advice, and constructive suggestions from the very beginning of this project. In particular, I thank Giacomo for his love and support over the years. His scholarly advice and generosity of spirit in working with junior scholars (myself included) is unmatched and deeply appreciated. Kathleen is the best friend that one could hope for. She provided support and encouragement through very difficult times; she is an inspiration as a scholar, friend, and mentor.

I thank Joseph S. Nye, Jr., for his support and interest in my research as well as Derek Reveron, James Goldgeier, Becca McBride, and Cameron Thies for their meticulous reading of my manuscript and their detailed comments on it. Numerous people offered helpful suggestions and advice during various phases of my research project: Clay Benton, Geoff Dancy, Scott DeMarchi, Christopher Gelpi, Hein Goemans, Joseph Grieco, Siegfried Hecker, Rick Herrmann, Ted Hopf, Patrick James, Jean-Marie Kamatali, Charles Krupnick, David Manning, Khalil Marrar, Jargalsaikhan Mendee, Jonathan Morgenstein, Steven Poe, Brian Pollins, Richard Price, Joseph Ralston, Scott Sagan, Brian Selmeski, David Sorenson, Joshua Spero, Alex Wendt, Steven Wilkinson, and Jenifer Whitten Woodring.

For their help with historical data and local resources I thank Aaron Higby and the Archives and Special Collections librarians at the Combined Arms Research Library at Fort Leavenworth, Kansas; research librarians at the Air University Library at Maxwell Air Force Base, Alabama; and the research librarians at the Library of the Marine Corps in Quantico, Virginia.

The Smith Richardson Foundation provided financial and intellectual support that was essential. I am grateful for their support for my project that extended over several years. I also thank the Mershon Center at Ohio State University, the Center for International Studies at the University of Southern California, and the Center for International Security and Cooperation at Stanford University, which all generously extended post-doctoral research scholarships to me. For their financial support of future projects based on the research themes presented in this book, I thank the Bureau of Educational and Cultural Affairs of the U.S. Department of State, the Council for International Exchange of Scholars, and the Bulgarian-American Fulbright Commission. I completed the final revisions of this project while at the Defense Advanced Research Institute at the G. S. Rakovski National Defense Academy (Sofia, Bulgaria), and I thank Yantsislav Yanakiev and members of the academy for their sponsorship, hospitality, and advice during my time in Bulgaria.

Finally, I am grateful to my wonderful editor, Marie-Claire Antoine, and her colleague Monica Savaglia for their supervision of my manuscript throughout the publication process. I deeply appreciate Marie-Claire's enthusiastic support for the project.

1

Military Soft Power in American Foreign Policy

In early 2011 successive waves of democratic protests swept across North Africa and onto the Arabian Peninsula. Quickly named "the Arab Spring," the movement included most countries in North Africa and on the Arabian Peninsula. Outcomes varied and included full-fledged democratic transitions, economic concessions, reform of government institutions, and violent suppression of protesters. Longtime dictators Ben Ali (Tunisia), Hosni Mubarak (Egypt), Muammar Qaddafi (Libya), and Ali Abdullah Saleh (Yemen) were deposed. To date, democratic elections have been held in Tunisia, Egypt, and in Morocco where King Mohammed VI quickly instituted a number of reforms. In war-weary Algeria the government repealed the state-of-emergency laws that had been in existence since 1991. In Bahrain protesters were violently suppressed with the help of foreign troops from neighboring Gulf Cooperation Council countries Saudi Arabia and the United Arab Emirates.

Seasoned observers were surprised by these events; first, for their appearance that no one had predicted; second, for their contagion effect in countries previously thought inured to both protests and democracy; and, third, for the success of a few of these protests in throwing out longtime dictators and bringing in democratic institutions and practices to replace the dictatorships. In each country the actions of the military as an institution and the decisions of individual military commanders were important factors in how the protests played out. In Tunisia, the military was seen as the protector of the people. As the protests were ongoing in Tunisia, David Kirkpatrick (2011b) of the *New York Times* reported protesters chanting: "Long live the people, long live the military" because the military was seen as protecting the people from the internal security militias of the Ben Ali dictatorship. In a similar story in Egypt, the Egyptian military eventually sided with the protesters and pledged not to use force against the Egyptian people. But events unfurled differently for the people in Libya and Syria. In Libya the military had historically been held at arm's length by Qaddafi and competed with many varied internal militias. This ambiguous relationship ended in 1986 when the United States conducted air strikes against Libya (almost killing Qaddafi) and the military gained status as the only effective force to protect the regime (Kamrava 2000: 85–87). The Libyan military continued in this role in 2011. While some elements of the army joined the protesters, Qaddafi's elite military units remained "unwavering until the very end, and they responded without restraint to the uprisings" (Lutterbeck 2012: 14). In Syria the government used (and is still using) substantial military force including the bombing of cities to suppress protests and retain power.

These democratic revolutions of the Arab Spring illustrate how the military plays a key role in democratic transitions—not only in subverting democratic reforms but also in creating the conditions for the success of democratic movements. As David Sorenson (2010: 137) so aptly put it: "The path to democratization in the Middle East [or elsewhere in the world] often runs through the military barracks." In the Tunisian Arab Spring the military laid the foundation for success of the democratic uprising. Army Chief of Staff General Rashid Ammar, Tunisia's top military officer, refused orders of the Ben Ali government to fire on the demonstrators and used his soldiers to protect demonstrators from Ben Ali's security and police forces. Kirkpatrick (2011a) reported that General Ammar was believed to have helped in efforts to convince Ben Ali to leave Tunisia, a key factor in bringing greater calm to the democratic transition. The Tunisian military's actions during the Arab Spring not only contributed to the success of the protests but also, as Cook (2011) has argued, are likely to have longer-term positive consequences for the nature of governance in Tunisia because "the military's decision to side with the people will likely put pressure on Tunisia's interim leaders and those who want eventually to come to power to heed society's demands for reform and change."

The democratic protests that swept across the Arab world were both unprecedented and unpredicted. In astute commentary on the nature of the Arab Spring, former Canadian ambassador Jeremy Kinsman[1] (2011: 38) wondered why those of us who are already "democratized" should have found the Arab Spring so surprising: "A major question for democracies is why it all came as a surprise. It always does. There have been 60 democratic uprisings since Portugal in 1974 (323 revolutions since 1900) and we are always surprised."

Before 2010 anyone predicting that a full-blown democratic revolution would sweep across a region with some of the most enduring and hardcore dictatorships would have been thought, at best, uninformed.[2] And certainly, anyone predicting that the militaries of some of these dictatorships would be open to and supportive of the democratic movements would have been thought totally out of touch with reality. In the Egyptian case, it was said that Egypt's military was simply too closely tied to the Mubarak regime to let it fail. Yet, the revolutions happened. As Hamid (2011) insightfully noted, "What has happened since January [2011] disproves longstanding assumptions about how democracies can—and should—emerge in the Arab World.

Even neoconservatives, who seemed passionately attached to the notion of democratic revolution, told us this would be a generational struggle. Arabs were asked to be patient, and to wait. In order to move toward democracy, they would first have to build a secular middle class, reach a certain level of economic growth, and somehow, foster a democratic culture. It was never quite explained how a democratic culture could emerge under dictatorship."

The lack of a viable explanation was in part attributed to U.S. policymakers' experiences in wars in Vietnam, Iraq, and Afghanistan where they had found out just how difficult it could be to instill democratic governance through the use of military force. In both Iraq and Afghanistan, the United States eventually designed more effective strategies that integrated the military's coercive force (hard power) with strategies to persuade and co-opt (soft power), alternatively known as "winning hearts and minds." Within political science the lack of a viable explanation was in part a result of how military power was conceptualized and studied. International relations scholars have largely focused on a top-down approach where military organizations are seen as instruments of the state used by the state to coerce and impose its will, usually on other states.

This book offers an alternative bottom-up view of how military organizations can influence political processes and decisions through the development of cross-border communities of military professionals. These worldwide military networks are formed and sustained, in part, through military educational exchange programs hosted by the United States' elite military schools, its war and staff colleges. In turn, the networks are one way that the United States extends its influence or "soft power" worldwide.

1.1. MILITARY SOFT POWER

Soft power, according to Joseph Nye (2011: 16) is "the ability to get preferred outcomes through co-optive means of agenda setting, persuasion, and attraction." Soft power is built through agentive strategies in the form of public diplomacy actions undertaken by diplomats to engage foreign publics (Nye 2011: 17). These include holding town meetings, giving interviews, organizing outreach programs, sponsoring international exchanges, and managing virtual communications (Clinton 2010). Soft power can also be gained through what Nye (2011: 17) calls structural effects, meaning setting an example that others wish to emulate—in short to serve as the "shining city on the hill."

Soft power is usually ascribed to public diplomacy efforts that are conducted by the U.S. State Department. But, other agencies of government also engage foreign publics and serve as examples that others might wish to emulate. The U.S. military is an interesting case when considering how soft power is built because U.S. military officers are usually associated with hard power, but they also frequently interact with their colleagues from other countries, for example, in interactions in alliances such as NATO. Thus, the U.S. military is in a position to serve as a builder of soft power and has incentives to do so.

This book explores one way the U.S. military builds soft power: U.S.-hosted military educational exchange programs; and the book focuses on one specific foreign policy impact: how a democratic culture can be nurtured within militaries around the world by the United States through these exchange programs. Successive U.S. Secretaries of State have heralded student exchange programs as a powerful source of soft power (Kerry 2013; Clinton 2010: 16; Rice 2006). John Kerry (2013) in his first official speech as U.S. Secretary of State argued that exchange programs "enable the most talented citizens to share their devotion to diplomacy and peace, their hopes, their friendships, and the belief that all of the Earth's sons and daughters ought to have the opportunity to lift themselves up."

U.S.-hosted military exchanges are important venues where military officers from different countries build social and professional networks. The friendships and professional contacts between foreign officers studying in the United States and their U.S. counterparts certainly help to improve the ability of the U.S. military to work with allied nations as well as potential coalition partners; in addition, these personal and professional military networks can support the development of democratic norms and practices through agentive and structural processes. This does ***not*** mean that military educational exchanges "cause" democracy. This is no more the case than saying economic development "causes" democracy or military expenditures "cause" autocracy. But, what it does mean is that when people who are repressed by autocratic regimes decide to take matters into their own hands by protesting and demanding democratic reforms, they are more likely to succeed because their own militaries are less likely to help repressive regimes stay in power. Additionally, when political leaders seek to build democratic governance, the exchanges are one way that they can expose their own military officers to civil-military relations in a mature democracy. As Margaret Mead (1942: 20) pointed out as America entered World War II, "Democratic procedures are not something that people have, like automobiles or hot-dog stands or a way of building roads"; it is a way of behaving and an attitude of mind that must be nurtured by society.

The presence of international officers at the United States' premier military educational institutions is substantial; yet, there is virtually no systematic empirical evaluation of the important long-term impacts that these exchange programs might have. For the United States, the goal of the military exchanges is multifaceted. Certainly they help to build relationships with other militaries that can improve the effectiveness of multinational operations. Just as importantly the exchanges serve as a source of information for participants about military topics, but also about the United States, U.S. people, and U.S. values. When the officers' experiences in the United States are positive, the exchanges help the United States to extend its influence through "soft" rather than "hard" power. This, in turn, helps the United States to accomplish its foreign policy goals, of which encouraging and supporting the development of democratic governance around the world is one of the most important.

Building military soft power is a process that takes place over numerous years, even decades, during which both the beliefs of individuals and the ideas espoused by their military communities are mutually constituted through many, many interactions. This book examines micro-level processes, meaning the motivations and attitudes of individual people whose ideas form the essence of their local military communities; as well as macro-level processes, meaning how U.S.-hosted military exchange programs affect political identity, behavior, and preferences of countries across the international arena in the longer term. In the book, these processes are broken into two analytic steps. In the first step, individuals (military officers) are socialized by larger political units (the United States through its war and staff colleges). In the second step, larger political units (the home countries of the officers) are socialized when members of these political units (military officers) bring new ideas, beliefs, and practices home with them.

In the first analytic step, the book examines how military soft power is built through U.S.-hosted military exchanges that socialize their international participants to particular beliefs and ideas. In fact, the overt stated goal of the exchanges is to do just that: to teach soldiers from around the world the ideas that they need in order to work effectively with U.S military forces. These ideas include concepts fundamental to U.S. military equipment, U.S. military doctrine, and U.S. operating procedures. Instilling the belief or persuading the learner that these ideas work well and are worthy of adoption is also an important aspect in building soft power. Another overt stated goal of the exchanges is to teach international participants about U.S. democratic values, such as the rule of law and internationally recognized human rights, and to expose the participants to how democratic institutions function in the United States. In addition to teaching concepts, the purpose of the U.S. military exchanges is to demonstrate the value of democratic institutions and democratic practices and to show that they are worth pursuing.

In the book, survey data from recent graduating classes at U.S. military war and staff colleges are presented to show what the officers learned in their programs, how they learned it, and then to assess the extent to which this achieves the programs' stated goals. The goal is to understand what types of ideas and perceptions of the United States foreign officers take home when their exchange program in the United States is over. These new ideas might be quite varied. Because the officers have just spent a year in the United States learning about U.S. military doctrine and U.S. military operating procedures, one set of ideas may result in changes in their own military institutions and practices. Also returning home with the officers are their personal perceptions of the United States, U.S. citizens, U.S. civil-military relations, U.S. government, and U.S. foreign policy. When those perceptions are positive, they help to build goodwill toward the United States as well as U.S. military soft power.

In the second analytic step, the book examines the longer-term impact of the exchanges on countries worldwide. Historical data is used to evaluate the extent to which the growth in democratic institutions (such as political parties and elections) and democratic practices (such as freedom of speech and rule of law) are associated with participation in military exchanges. As Margaret Mead (1942) argued, democracy is a mindset and way of life, not a material object. This mindset and accompanying way of life is absent in authoritarian countries, and weak in transitioning countries such as ex-Soviet client states, thus, it must be learned and nurtured. Successful transitions to democratic governance depend on citizens' knowledge about how democratic institutions should work in their everyday lives. One way to gain this knowledge is by living for an extended period in a mature democratic country.

1.2. U.S. MILITARY ENGAGEMENT ACTIVITIES

The research presented in this book began with observations and anecdotes told by U.S. military officers and their international colleagues.[3] To date, most of these observations have remained just that, personal anecdotes about the friendships formed and the bonds maintained between international officer students and their counterparts from both the United States and other countries. These anecdotes routinely emphasize how friendships between military officers from different countries have endured over many years. This experience is widely shared by many of the U.S. military officers who have attended the programs at the U.S. military's war and staff colleges. This shared experience is not trivial in scope. Each year hundreds of international officers come to study at the U.S. military's most elite schools. In 1994, the U.S. Army Command and General Staff College celebrated one hundred years of hosting international officers. As of 2012, more than 7,300 international students have attended the U.S. Army Command and General Staff College, just one school within the U.S. military's education network. Of these, 241 have become heads of their militaries or heads of state (CGSC *Hall of Fame* 2011). Despite these significant achievements by the Command and General Staff College's international graduates, Stewart (2010: 81) in his study of officer education at the college was surprised to find that "historians have paid scant attention to the participation of international students" there.

Historically, educational exchange programs in general did not have many participants in their earliest years in the United States because U.S. universities and their curricula were seen as inferior to their European counterparts. This perception began to change as U.S. power, military might, and economic influence grew in the aftermath of World War I and later World War II. One of the most important early developments in U.S. civilian educational exchange programs was the founding of the Institute of International Education (IIE) in 1919. IIE's founders (including two Nobel Prize laureates) believed that international educational exchanges were the single most effective way to build understanding across the globe and in so doing promote cross-border understanding and lasting peace. Initially, IIE programs sponsored European scholars, but in the 1930s these programs expanded to include scholars from the Soviet Union and Latin America (IIE 2011).

During the time that U.S. civilian educational exchanges were gaining ground in the United States, there were only a very few participants in U.S-hosted military educational exchanges. This changed in the late 1930s as the U.S. government sought to counter the growing presence and influence of German, French, and Italian training missions in Latin America. According to a post-war study conducted by the Air Historical Office, from 1941–1945 the U.S. Army Air Forces provided technical and flying training to over 18,000 students from thirty-one countries. Roughly half of the students were British. Other major participating countries included France, China, Netherlands, Brazil, Mexico, Turkey, Yugoslavia, and many Latin American countries (USAAF 1947: 1–5). The exchanges provided the foreign participants familiarity with U.S. doctrine, operating procedures, and equipment. However, for the U.S. government, a major goal was to extend its influence or soft power within other countries, as the post-war study noted: "The United States hoped to profit from the feeling of familiarity with and friendship for the United States developed in the foreign officers trained" (USAAF 1947: 1).

Despite cooperative efforts to work together in alliances and to learn from each other, military organizations are stereotypically seen as very insular organizations encompassed within the territory of a single home country. But, this is not how many modern militaries operate; certainly it is not true of the U.S. military. Since the end of World War II the United States has sought to place its military at the center of a web of military-to-military interactions where U.S. military personnel routinely interact with their counterparts in other countries. This presence was initially focused on ensuring that past foes Japan and Germany were brought into the American sphere, rebuilt, and democratized. To support this mission, 136,554 U.S. troops were stationed in Japan and 97,820 in Germany in 1950. While the bulk of U.S. troops were stationed in these two theaters of operation, U.S. troops were a permanent presence throughout the world, with a little over 22% of all U.S. military personnel stationed overseas (DoD *Personnel Strengths* 2010). Sixty years later the U.S. military still maintains a significant presence spread widely across all regions of the world with U.S. military personnel stationed in 160 countries. At the end of 2012, there were at least 37 countries in which there were 50 or more U.S. military personnel; 12 countries with more than 1,000 U.S. military personnel; and six countries (Afghanistan, Japan, Germany, South Korea, Kuwait, and Italy) with more than 10,000 U.S. military personnel stationed on their soil (DoD *Total Military Personnel* 2013). There are also countries such as Bulgaria that do not have a large permanent U.S. military presence but seek to establish one.[4] As a proportion of U.S. military strength, the overseas presence of U.S. military troops is substantial. Roughly 24% of all U.S. military personnel were stationed outside of the United States in 2012 (DoD *Total Military Personnel* 2013).

Not only are extensive numbers of U.S. soldiers stationed abroad, but also many foreign military personnel are stationed in the United States as part of diplomatic corps as well as within U.S. military commands. While most are incorporated into

U.S. military organizations, there are a few countries that have their own military units stationed on U.S. soil. For example, the German Luftwaffe maintains its own flying training center at Holloman Air Force Base in New Mexico with roughly 600 German military personnel permanently stationed with this training center (U.S. Air Force 49th Fighter Wing 2010). The U.S. Air Force recently announced that the Royal Saudi Air Force would maintain its own maintenance and training unit at Mountain Home Air Force Base in Idaho. Roughly 150–250 Saudi Arabian military personnel will be stationed there beginning in 2014 (Kok 2010).

In addition to the stationing of military personnel at overseas locations as part of operational units, military personnel also attend foreign military schools and other educational forums. For example, each year there are thousands of foreign military soldiers who attend conferences, war games, training courses, and military colleges within the United States. Roughly 7,000 foreign military and civilian officials attend formal training in over 4,000 courses at approximately 150 military schools and installations across the United States each year (U.S. Department of State, Bureau of Political-Military Affairs 2008). These officials come from around the world: over 140 nations routinely send military students to the United States for training. The U.S. State Department argues that these programs "enhance the ability of friends and allies to participate in coalition, humanitarian, peacekeeping, counterterrorism, and counterinsurgency operations. Military assistance also provides a valuable means of engaging with foreign militaries on issues such as civilian-military relations and respect for human rights" (DoS *Congressional Budget Justification* 2010: 187).

1.3. MILITARY EDUCATIONAL EXCHANGE PROGRAMS AT U.S. WAR AND STAFF COLLEGES

At the United States elite military schools, the presence of foreign officers is substantial; they make up approximately 10–20% of the student bodies at U.S. war and staff colleges. The U.S. Army War College plans to expand that number to 20–25% in the coming years. Many of the graduates have become heads of their militaries or occupy important political positions. For example, from 1957 to 1994, over half (53%) of the international students who graduated from the U.S. Naval Command College achieved flag rank (the naval equivalent of general officer rank), and 19% of these went on to become the chief or highest ranking officer of their service. Likewise for the U.S. Army War College during the same time period, 54% of its international officer graduates achieved general officer rank. The presence of these international graduates is worldwide: these general officers came from eighty-five different countries (Cope 1995: 24–25). The U.S. Department of State has reported that during 2009, 1,549 military personnel who had attended training in the United States occupied national-level leadership positions in their home countries (DoS *Congressional Budget Justification* 2010: 282). The exchanges at U.S. war and staff colleges receive the bulk of Congressional funding that is devoted to all of the various types of foreign military training programs (U.S. GAO 2011: 4).

The officers that come to train in the United States are not just from those countries considered traditional U.S. allies, such as Great Britain or Germany, but are from a vast variety of countries. They come from democratic countries and non-democratic countries, from wealthy countries and those struggling with economic development, and from traditional U.S. allies and emerging partner nations. The potential influence of the international graduates of U.S. war and staff colleges is significant. The U.S. Department of State argues that as the personnel who have attended schools in the United States attain national-level positions, "the skills and values provided in that training will eventually be spread to the entire military structure, and the leadership will be more likely to respect civilian control of the military, be willing to work with U.S. led or sponsored peacekeeping missions, and be interested in maintaining a longstanding relationship with the United States" (DoS *Congressional Budget Justification* 2010: 282).

1.4. TRACKING LONG-TERM IMPACTS

The presence of international officers in the United States' premier military educational institutions is substantial, and the goals of these exchanges are explicit, yet there is little systematic empirical data on the exchanges' long-term impact on international politics or U.S. military operations. In the last decade, the U.S. Department of Defense has increased its ability to track the careers of the graduates of the war and staff colleges, but there is still much work to be done. The U.S. Government Accountability Office (GAO) has repeatedly criticized both the Department of State and the Department of Defense for inadequate tracking of military exchange participants and lack of established procedures to assess long-term program performance in accomplishing its stated goals (U.S. GAO 2011: 1, 16–17). While the tracking of the graduates of all international military and education exchange programs is limited, more attention has been focused on the higher ranking graduates of the professional military schools. Even for these higher ranking graduates, however, monitoring has been uneven once the graduates return to their home countries. While some U.S. in-country training managers have been able to maintain career information for many graduates, other managers have been less successful. Problems have included lack of resources and personnel to collect data, negative reactions by foreign governments to U.S. attempts to monitor their officials, and lack of clear guidance for the training managers (U.S. GAO 2011: 17–21). Even for those training managers who had been able to collect and maintain substantial information, this information has not been systematically shared with higher echelons in the Department of Defense or Department of State (U.S. GAO 2011: 19), resulting in difficulty assessing overall program performance systematically and across time.

Former Vice Chairman of the Joint Chiefs of Staff General Joseph Ralston has pointed out that a key issue in tracking the graduates of U.S. military war and staff colleges is that there is no central office within the Department of Defense or the Department of State that manages or maintains data on the accomplishments of international military students (Ralston 2010). Rather than a centralized administration of the international military exchanges, each individual war college or staff college manages and runs its own program for international officers and tracks the successes of its graduates. My discussions with faculty members at war and staff colleges also confirmed this key observation. For the most part, the interactions between personnel associated with the various international officer programs that do occur are generated by faculty members themselves who reach out to other organizations, rather than any sort of official top-down policy or oversight organization.

Individually, the professional military schools have recently begun to institute programs that might provide better contact with their international graduates; from alumni websites to the occasional reunion. Although the individual schools have tried to track their international graduates with varying levels of effort and varying levels of success, there have only been limited and largely unsuccessful attempts to compile comprehensive historical data. A former staff member of the U.S. House of Representatives International Relations Committee who had been tasked in 1978 with assessing the effectiveness of the training of foreign military officers remarked to me in 2007 that "no one knew whether or not the training was having any impact as desired by the Congress. The Army and Air Force at least kept records on the kind of training provided to foreign military personnel, be it under grant or contract auspices. The Navy kept no such records."[5] As an example, in the 2011 U.S. GAO report cited above, the GAO used data from fiscal year 2000 onward because earlier data were deemed unreliable (U.S. GAO 2011: 2).

One of the most comprehensive reviews of the military exchanges was conducted by the U.S. National Defense University in the mid-1990s and headed by John Cope. Cope and his researchers found that efforts by either the U.S. military or the U.S. State Department to maintain contact with foreign graduates had been "ad hoc and highly personalized" (Cope 1995: 46). Their study noted:

> [U.S.] Embassies are irregular in keeping information on students trained under the IMET [International Military Education and Training] or FMS [Foreign Military Sales] programs—recordkeeping too often depends on individual motivation. The INSS [Institute for National Strategic Studies] team found no record of embassy initiatives to keep in contact with officer-graduates in a manner similar to that accomplished for other foreign education programs, such as the Fulbright program. DSAA [Defense Security Assistance Agency] does not have complete or adequate historical records on IMET graduates other than an annual report on positions of prominence (undefined) attained by former students. There are no data on officers educated with FMS funding. Also incomplete is information on the accomplishments of graduates. (Cope 1995: 46)

This situation had not improved fifteen years later. In 2010, Admiral Willard, the Commander of U.S. Pacific Command, testified before the U.S. Senate Armed

Services Committee that security assistance programs in the Pacific region had "not evolved much since the end of the Cold War" and were constrained by a patchwork of authorities, persistent shortfalls in resources, and unwieldy processes (Willard 2010: 5). Unlike civilian exchanges, such as the Fulbright program, the U.S. government has made only sporadic efforts to track international military officer graduates. Because of this, there has been little evidence accumulated and few mechanisms developed to evaluate the important impacts that the military exchanges might have on individual participants, on the ability of participating countries' militaries to operate effectively with U.S. forces, or on domestic politics in the home countries of the military officers.

U.S.-hosted military educational exchange programs have received even less attention within political science scholarship that has focused on the hard-power aspects of the military. International relations scholars have extensively examined how military forces are used to coerce, deter, and fight adversaries, but have virtually ignored the social and professional networks that exist between soldiers of different countries. Within the broader discipline, scholars Michael Barnett and Raymond Duvall (2005: 41) argue that "the failure to develop alternative conceptualizations of power [has limited] the ability of international relations scholars to understand how global outcomes are produced and how actors are differentially enabled and constrained to determine their fates."

Understanding the motivations and preferences of military and political leaders has been most widely pursued by civil-military relations scholars who have sought to show how military personnel could best be subjugated to the will and preferences of civilian political elites. While these studies are important, they are also focused on the internal dynamics within single countries rather than the U.S.-centered international networks examined in this book. Scholars of civil-military relations have largely sought to explain how various facets of power are contested and shared between military and civilian elites and to advocate for particular power sharing arrangements most conducive to civilian control over military power.[6] These scholars have contributed to our understanding of the political processes and dynamics within countries. They address important questions of how norms, values, interests, and behaviors are formed and acted upon within domestic politics of countries. However, their focus has been primarily on civil-military dynamics within countries.

In the international realm, scholars have almost exclusively studied the military as an instrument of force and coercion, not as a mechanism of soft power where new ideas are learned and subsequently impact behavior. In reference to international relations scholarship in general, Peter Haas (1992: 2) argued that "in focusing on the structure of international or domestic power in their explanations of policy coordination, many authors ignore the possibility that actors can learn new patterns of reasoning and may consequently begin to pursue new state interests. While others mention this possibility, few investigate the conditions that foster a change in state interests and the mechanisms through which the new interests can be realized."

The following chapters address the challenges described above and serve as a foundation for further research. The book shows that participation in educational programs hosted by the U.S. military's war and staff colleges for international participants is one way that important and influential people learn new patterns of reasoning, acquire new sources of information, and consequently impact their home country's domestic politics and its actions in the international arena. By focusing on how ideas are transferred through international military networks, the following pages provide novel insights into how military organizations influence international politics and civil-military relationships.

1.5. PLAN OF THE BOOK

The book is organized into three parts along with this introductory chapter and a concluding chapter. Chapter 1 has provided information on the importance, nature, and scope of educational exchanges hosted by the U.S. military as well as the extent to which the officers who have participated in these programs have attained national-level positions in their home countries. The chapter introduced the book's central argument of how U.S. military war and staff colleges help military officers from around the world acquire beliefs and perceptions of the United States that can impact their future actions. It laid out the theoretical framework for how international politics might be shaped by the United States through social and professional interactions between soldiers of different countries within a U.S.-centric international community of military officers.

The chapters that follow are organized into three parts. Part I (chapters 2 and 3) is the theoretical core of the book; it examines how military organizations can serve as effective agents of political socialization and contains two chapters that provide detailed theoretical and historical information. The book's theoretical argument integrates ideas from four areas of research: (1) international relations theories of military power, (2) soft power theories of how influence can be obtained through social interactions, (3) constructivist theories of how epistemic communities serve as transnational channels of learning and influence, and (4) social theories of how educational exchanges serve as socializing environments.

Chapter 2 defines and discusses processes of socialization. It emphasizes that social interaction, identity, and learning are essential, but understudied, aspects of socialization. The chapter begins by examining the ways that international relations scholars have approached the study of political socialization primarily as a behavioral outcome of an asocial rational decision-making process where the state is the primary unit of analysis. It then explains why a new approach is needed. The book's new model of political socialization focuses on the socialization of individuals and their communities as a long-term continuous process of mutual constitution in which more attention is paid to how individual people are socialized and the impact of this process on the state. The model incorporates the role played by communities, both formal and informal, and shows how they shape their members' core views and motivate the actions that members take. In turn, members bring new ideas into these communities that impact the roles, missions, and self-image that the community espouses. The chapter concludes with an explanation of why and how military educational exchanges are a particularly important environment for political socialization.

The book's theoretical argument is developed further in chapter 3 with a discussion of how attendance at U.S. military schools socializes participants and how this process builds U.S. soft power and influence. The chapter examines two conceptualizations of U.S. military power: the traditional rationalist view that emphasizes material capabilities and their instrumental use, and a nontraditional constructivist view that emphasizes the impact of common identity and shared ideas. The purpose of this discussion is to show how different conceptualizations of military power have shaped the study of military organizations and the processes through which they exert influence on the international system. As a key part of the book, discussion of the second nontraditional view advances the study of political socialization as a means of understanding how military organizations acquire soft power. It argues that the United States exerts political influence as part of an international military community in which military exchanges play an important role in socializing others. For U.S. foreign policy this military community is important because it helps to build goodwill toward the United States and in the process creates U.S. military soft power. The main claim is that in addition to their function of fighting wars, deterring conflict, and coercing adversaries, the United States military also exerts soft power. The chapter closes with a description of a number of U.S. military programs and activities that promote the development of a U.S.-centric international community of military officers.

Part II (chapters 4, 5, and 6) analyzes the impact of U.S. military war and staff colleges on the exchange officers and their families using original survey data. This part of the book examines the professional and personal experiences of the officers that underpin the larger international macro-level trends discussed later in part III. Chapter 4 begins the book's detailed focus on U.S.-sponsored military educational exchange programs. It first describes the U.S. government's centerpiece program called International Military Education and Training (IMET). IMET is the U.S. government's primary program that provides grant funding for a variety of education and training programs for foreign military personnel. The chapter discusses the nature, scope, goals, and costs of IMET programs and explains the extent to which the international exchanges at the war and staff colleges are encompassed under IMET. The chapter then looks specifically at the curriculum and international exchange programs at each of the U.S. military's war and staff colleges. Each school's international exchange program is described including its historical evolution, data on class sizes, the proportion of foreign attendees in the classes, and the extent of integration of the foreign officers with U.S. officers in the classes. The chapter also provides data on how many of the foreign graduates from each school have gone on to hold the high-level command positions in their country's armed forces or government.

Chapter 5 pries open the "black box" of the military in order to look at the people who constitute military organizations. The analyses in this chapter are based on my original attitudinal surveys of foreign officers at U.S. military war and staff colleges. The chapter uses the model of socialization discussed earlier in the book in order to assess socialization in actual practice and from the view of the participants in the exchange programs. The chapter begins with a description of the survey; how and where the survey was conducted and basic demographic statistics on its participants. The chapter provides data on the backgrounds of the international officers who participated in the survey, the regions of the world that they were from, their promotion potential, the status of democracy in their home countries, their previous international experience, and the extent to which they were integrated into U.S. social and professional communities during their time in the United States. Unlike young high school or college students who go abroad on exchanges, the international officers have substantial ideological and professional ties to their home governments, they view their American experience as a tool to advance their careers, and thus they should be relatively impervious to socialization attempts. The surveys show how the officers were extensively integrated, both socially and professionally, into their U.S. military program, U.S. communities, and U.S. society. The outcome is that this experience then led to more positive views of the United States and greater understanding of the values that the United States espouses.

Chapter 6 continues to meld theory with practice and shows how soft power can be built through person-to-person interactions. The survey data give voice to the many international officers who explain from their own points of view what knowledge, beliefs, attitudes, and perceptions they acquired during their time in the United States. A key finding from this chapter is that the officers are much like international exchange participants the world over. They come away from their time abroad with increased understanding and warm feelings toward their host country and the people who were their hosts. This was true for officers from authoritarian countries as well as those from democratic countries. This chapter demonstrates that the military exchange programs are a source of soft power for the United States, a claim often made but hardly ever demonstrated.

Part III (chapters 7 and 8) uses quantitative modeling techniques to assess the long-term impact of U.S. military soft power on international trends in democratization. In chapter 7, hazard models show that the military exchanges hosted by the U.S. military's war and staff colleges were positively associated with the development of democratic institutions across time and space. Countries that participated in the military exchanges were more than two times as likely to develop more democratic institutions than those countries that did not participate. The effect was most pronounced for "hardcore" autocratic countries. In chapter 8, ordered logit models with random intercepts show that military exchanges were associated with improved democratic practices in the countries that participated. Participation in military exchanges increased the likelihood that a government would respect its citizens' rights for basic freedoms (such as the freedom of movement, speech, religion, political participation, and workers' rights) by nearly 8%. The findings in chapters 7 and 8 obtain while controlling for a large array of pre-treatment confounders, meaning potential factors that influence participation in the exchange programs as well as the outcomes on the dependent variables.

In chapter 9, the concluding chapter, the book's theoretical contribution to the study of political socialization, soft power, and military organizations is summarized. The chapter discusses significant policy implications of the book's findings to include the importance of military exchange programs as a mechanism of soft power, the importance of engaging elites in nondemocratic countries rather than excluding them from U.S. programs, and ways that U.S. policymakers can support and strengthen the military exchange programs as an instrument of U.S. soft power.

The arguments and findings in this book provide new insights into how military institutions and their personnel influence *international* politics beyond simply being used as an instrument of coercion or conceived of as a collection of material resources. As Haas argued, individuals can and do learn new patterns of reasoning. The chapters that follow provide one theory of how politically relevant individuals might learn new ideas and show how this process can have a significant impact on the domestic politics within their individual countries as well as shape how those countries engage within the international arena. The analyses highlight the important role that military soft power can play in the broader context of U.S. foreign policy.

NOTES

1. Jeremy Kinsman, one of Canada's most experienced diplomats, has served as ambassador to Moscow, Rome, and the European Union in addition to other assignments.

2. Kinsman (2011: 39) reported that in U.S. President Obama's first National Security Council meeting that discussed the possibility of the Tunisian revolution spreading to Egypt, his regional experts gave it only a 20% chance of happening.

3. For an excellent study along these lines the reader should investigate David F. Manning's ongoing multivolume Global Arms series that provides interviews and testimonies from military officers from around the world on how common bonds are formed and endure between them and their U.S. counterparts through the exchange programs hosted by the U.S. military.

4. In December 2012 the Bulgarian defense minister requested that the United States establish a permanent U.S. military presence at one of Bulgaria's training bases.

5. Interview with the author; not for attribution.

6. See Feaver (1999) for a comprehensive review of political science-oriented civil-military relations scholarship as well as Sorenson (forthcoming) for insightful analyses of civil-military relations in the Middle East.

I

THEORY AND ARGUMENT

2

Political Socialization and Educational Exchanges

Research on political socialization encompasses a broad range of disciplines, topics, and subjects. It seems simplistic to say that social interactions are key to socialization, but within the discipline of international relations, political socialization has often been studied as an asocial rational process in which an actor weighs costs and benefits, considers his own preferences, and then decides on his course of action. This asocial decision-making process has little to do with the actor's obligations as a member of a society and the concomitant responsibilities and collective beliefs the individual bears as a member. Yet, these are key issues in understanding how soft power can be built through effective public diplomacy programs such as educational exchange programs. This chapter will review key factors that international relations scholars have identified as important to the political socialization of individuals and states. It will explain why educational exchanges are effective instruments of political socialization and how U.S. military soft power results. In order to place this discussion within the context of the book's focus on identity and social interactions, the chapter begins by defining socialization from this perspective.

2.1. SOCIALIZATION

Socialization is a process in which people are exposed to new information and then incorporate this knowledge into how they see themselves, their world, and their place within that world. Traditional definitions of socialization focus on how a person, usually a novice, becomes part of a wider community through interactions with the community's members. For example, sociologists Peter Berger and Thomas Luckmann (1966: 130) defined socialization as "the comprehensive and consistent induction of an individual into the objective world of a society or sector of it." In the words of international relations scholar Alastair Johnston (2001: 494), socialization "is a process by which social interaction leads novices to endorse 'expected ways of thinking, feeling, and acting' . . . socialization is aimed at creating membership in a society where the intersubjective understandings of the society become taken for granted."

According to Virginia Sapiro (2004: 2–3), research on socialization is usually framed in one of two ways. In the first approach, scholars have sought to understand the processes and patterns through which individuals learn about and develop their relationships with political communities. In the second approach, scholars have investigated how political units (such as states, international organizations, or other polities) inculcated values, norms, and practices into their members. Both of these approaches focus on the socialization of individuals, but not their communities. That is to say, the beliefs or identity of a community were assumed to remain constant over time, but the beliefs or identities of individuals were changed. Reflecting this same orientation, Dawson and Prewitt (1969: 13) explain that for individuals, political socialization is a learning process through which an individual acquires his political views, but for communities, political socialization is also a top-down process of cultural transmission and indoctrination as new members join the community and "acquire its established patterns of thought and action." For Dawson and Prewitt (1969: 13) these two perspectives were complementary: "studies should proceed at both levels, each type of study drawing on the findings of the other." In sum, most studies treat socialization as a process undergone by individuals but not their communities, which are only conceptualized as status quo agents of socialization rather than recipients.

In these studies, exposure to new information, participation in social activities, observing other members, and formal instruction are all important ways that new members are socialized into an existing community. The use of language, both written and spoken, is instrumental to inducting new members into an existing community. Political theorist Hanna Pitkin (1972: 106) explained, "As a layman is gradually trained to become a scientist, he is initiated into a new language, new ways of seeing, a new world." As one example of the role of language in socialization, research on communication and political socialization has looked at how communicative factors shape the beliefs and behavior of children within their families and within primary and secondary schools. Recently, scholars have broadened this line of research by examining the impact of news media on the political socialization of young adults (McLeod and Shah 2009).

Socialization involves both individual learning and the transmission of cultural understandings. As Pitkin (1972) pointed out, learning one's language is an intensely social activity. It is a process not only of learning words and grammar, but how to use this knowledge appropriately with one's community to express and understand not just the definitions of words but the meaning of what is said—or even more context dependent, the meaning of what is not said. The same dependence on context is true for how individuals acquire their political orientations. Each individual has a political self that consists of "his entire complex of orientations regarding the political world, including his views toward his own political role" (Dawson and Prewitt 1969: 17). These orientations are acquired within the context of the myriad of different communities to which people belong from families to schools to work centers.

The processes and factors affecting socialization discussed so far have all focused on how individuals are socialized by a larger community, but this is only half of the story. As described above, socialization is usually thought of as a process whereby individuals acquire the norms, values, and practices of a community. Researchers have sought to understand how individuals' ideas evolve and change based on membership within a group, but not on how a group's norms and rules change because they adopt new members. Yet, communities are also socialized as new individuals are incorporated into them. When new members join a group, they may in turn bring values, norms, and beliefs of their own into the group that may alter the group's nature. In this sense, socialization has to do with the contestation and constitution of identities whereby individuals learn and accept the beliefs, values, and rules of the communities of which they are members or aspire to be members. In turn, individuals bring new ideas to their communities that may alter the beliefs, values, and rules of the community. As Alexander Wendt (1999: 340) put it, although identities appear to be stable, they are "always in process, always contested" as people interact and exchange ideas about themselves and their world.

Recognizing these dual processes, the model of political socialization introduced in this book presents the socialization of individuals and their communities as a long-term continuous process of mutual constitution. On one level of analysis, individuals (military officers) might be socialized by larger political units (the United States and the military epistemic community) as they attend military schools in the United States. On another level of analysis, larger political units (the home countries of the officers) might be socialized when members of these political units (military officers) bring new ideas, beliefs, and practices into them.

In sum, socialization is a process of the forming and reforming of identities and beliefs. A community teaches its rules, values, and beliefs to new members who gradually incorporate these ideas into their own identity and belief systems. The socialized individual has experienced, learned, and accepted those teachings and as a result has become part of the community. Over time the community itself evolves because as new members are incorporated, they also bring some of their own ideas with them. Thus, the identity of the community also changes over time. It is useful to keep this conceptualization of socialization in mind because (as discussed in the next section) international relations scholars have had a very different approach to the study of political socialization.

2.2. POLITICAL SOCIALIZATION AS A FIELD OF STUDY

Political scientists have long been interested in understanding how people acquire their political outlooks and party affiliations. Political socialization as a field of study in the United States began in the early 1960s as scholars sought to understand how young children were socialized to political orientations such as citizenship, political ideologies, and affiliation with political parties. Over the decades the field did not diverge much from its original focus, but due to a number of problems "lost its energy and declined to near oblivion" (Hepburn 1995: 6) in the 1980s.[1] Within the discipline of international relations, research on political socialization did not emerge until well into the 1990s. Scholars G. John Ikenberry and Charles Kupchan (1990: 284) remarked upon the paucity of research on it at this time: "Socialization is an important element of power, but we have meager tools with which to understand the mechanisms and conditions of its operation."

The study of political socialization was resuscitated in the 1990s because it had a new mission: supporting (or at least attempting to explain) democratic political socialization in emerging democratic states of the former Warsaw Pact. When the Soviet Union disintegrated, there were suddenly many countries adopting very different governing systems from those they had under Soviet rule. After years under authoritarian governments, the government officials of the new democracies were struggling to build democratic institutions and function effectively within these institutions as well as help their citizens learn about their duties and responsibilities as democratic citizens. This generated an interest in learning how the development of democratic values and practices within the newly emerging democratic states could best be supported. U.S. and European policymakers were very keen to help the new democracies survive and designed programs, organizations, and policies to promote and support democratic institutions.[2] U.S. scholars were less interested, and the impetus for research on democratic political socialization came mainly from within universities in Europe. As U.S. scholars Richard Niemi and Mary Hepburn (1995: 13) noted, "In sharp contrast to the situation in some established democracies, concern among political scientists [in Central and Western European countries] about political education and research into programs intended to promote the learning of democratic knowledge and attitudes [were] viewed as significant if not urgent." With this new mission, political scientists as well as scholars and educators from other fields were soon engaged in research on the processes of democratic political education and democratic political socialization.

2.3. DEMOCRATIC POLITICAL SOCIALIZATION IN INTERNATIONAL RELATIONS THEORY

Political socialization is an emerging and eclectic field of study in international relations, as Johnston (2001: 489) noted: "Predominant IR theories either ignore the possibility of socialization or are unprepared or unwilling to theorize about it." For Kai Alderson (2001), the study of political socialization has lacked a unifying framework. He noted that international relations scholars use the term "state socialization" in a myriad of different ways and "there is no consensus on what it is, who it affects, or how it operates" (Alderson 2001: 416). Similarly, Sapiro (2004: 5–6) criticized the discipline's lack of knowledge: "We are limited in our understanding of the process of political socialization—what agents are involved, when, and how, and what moderates socialization and learning—and not just in our knowledge of what is learned in different places."

In the United States, studies of democratic political socialization have focused primarily on the role of international institutions or a hegemonic state in socializing and liberalizing nondemocratic states. It is a path that has spawned a large literature but lacks the individual touch; that is to say, scholars have paid very little attention to how individual adult human beings contribute to the socialization processes of the larger political units. In her review of recent U.S. scholarship, Sapiro found a lack of comparative, transnational, and historical work. The "more fundamental problem for the field of political socialization as a whole," Sapiro (2004: 5) wrote, "is that its basic questions, theories, and conclusions about how people become part of the fabric of their political communities and how they develop their political orientations and practices are framed by observations of an excessively limited range of political contexts."

International relations scholars use the word "socialization" to describe many different sorts of processes and outcomes. The majority of international relations research has examined how states are socialized to behave in a particular manner. The socialization of states is conceived of as a process through which a state's behavior is changed by external factors, but this behavioral change is not the result of a change in the internal nature of the state. For some scholars, the explanation for why states behave as they do is to be found in the lack of central authority in the international system, for others the explanation involves membership in an international institution or the role played by very powerful states. As described below, these explanations rely on two general sorts of arguments; one having to do with rationalist considerations based on material capabilities (such as economic or military resources used as incentives or coercion), and the other having to do with ideationally based processes (normative pressures, persuasive arguments, learning, and identity formation). There are two general types of outcomes of socialization: rationalists point to behavioral changes but not identity changes whereas other scholars seek to explain identity changes that then result in behavioral changes.

2.3.1. The Rationalist View of Socialization

In the United States, international relations was dominated by neorealist scholars throughout the Cold War. Socialization was considered a process whereby different countries all came to behave in a similar manner despite the fact that they might have quite different types of governments and ideologies. The military was seen as a unitary tool used by the state to coerce its adversaries. Neorealists argued that all states would seek to gain and maintain power relative to others in the international arena because those that did not would be marginalized or eliminated from the system. The nature of the state's identity did not matter; what mattered were its material resources. As Kenneth Waltz (1979), one of the most prominent scholars of this era, argued, "The theory says simply that if some do relatively well, others will emulate them or fall by the wayside" (Waltz 1979: 118). According to Waltz, states compared their material capabilities (primarily military weapons and manpower, but also economic resources that supported the development of military capabilities) with the capabilities of potential rivals and then sought to ensure that they had enough material resources not to be conquered or otherwise subjected to the rule of their rivals.

This neorealist behavioral dynamic was enabled by the lack of hierarchical order in world politics. Unlike states that had a central government that regulated how citizens behaved toward each other, in international politics there was no central authority to enforce order between states or compliance with agreements that they made with each other. States had to look out for their own interests and their own security by acquiring enough power resources to make sure others did not push them around. According to this school of thought, states with different sorts of internal characteristics and different governing systems were nevertheless "socialized" to behave in a similar manner, that is to acquire power and seek to balance the power of others. According to neorealists, it did not matter whether a state was a dictatorship or a democracy, it should behave in the same manner toward other states. If a state failed to conform, it was likely to be subjugated to the will of other states or even lose its independence. For neorealists such as Waltz (1979: 76), state socialization was a process that "limits and molds behavior" of the state rather than a process that changed the internal characteristics or identity of the state. It was a theory devoid of any discussion of individual human agency.

In recent years, scholars have sought to incorporate ideational mechanisms within the neorealist theoretical framework while retaining the key neorealist insights for how states behave in relation to each other. Cameron Thies (2012) is one of the leading scholars pursuing this effort.[3] He has drawn on role theory from sociology and social psychology to examine how states identify and adopt roles for themselves, a process he calls "role location." In his early work, Thies (2001: 708) identified four master roles that states might adopt: great power, major member, minor member, and emerging state, and he sought to elicit social aspects of rivalries between states. In this framework, socialization referred to how states choose and compete for roles within the international system conditioned on the distribution of material capabilities between them. In this sense, role theory takes into account the evolving identities of states as they take on and discard various roles. While role theory has added explanatory depth to neorealism, it remains, like neorealism, focused on how the international system "socializes" states and was not intended to encompass the processes whereby individual people serve as either agents or recipients of socialization.

One of the more interesting research programs on political socialization that spanned the Cold War and its aftermath was hegemonic socialization theory that sought to understand how the most powerful states socialized other states. Scholars looked at how a powerful state might exert influence on other states, through both material and ideational means. Representative of the importance placed on material resources and coercion during the Cold War, Robert Gilpin (1981) argued that the influence of a hegemonic state, such as the United States or the Soviet Union, rested primarily on its material capabilities. In his research, he examined three ways a hegemonic state could obtain its goals: the use of military and economic power to coerce, the ability to provide public goods, and the sharing of norms and beliefs between states. Much like Waltz and other international relations scholars writing during the Cold War era, Gilpin placed great emphasis on economic and military coercion and discounted the provision of public goods and shared beliefs as weak to ineffective mechanisms of influence. For some, this seemed understandable at a time when U.S. international relations scholars as well as U.S. policymakers viewed international politics largely as a competition for survival between the two superpowers. For others, there seemed a large gap in understanding how powerful states operated.

Neoliberal scholars in the United States have contributed to a large body of research that seeks to show how international institutions "socialize" the behavior of states by constraining their use of power and enabling them to work with each other. Neoliberals argue that state behavior might be "socialized" not by the international system per se as the neorealists argued but by introducing limited central authority within the international system in the form of international institutions.[4] According to this line of research, international institutions provide a forum within which states can be cajoled, pressured, or otherwise convinced by other member states to behave in the manner prescribed by the institution; that is to say, states are socialized to behave according to the norms and rules of the larger political unit.

Theories of how international institutions socialize state behavior include rational-instrumental explanations, ideational-normative explanations, and a combination of the two. For example, March and Olsen (1998) argued that international institutions linked states together through a series of norms, identities, and collective beliefs that can alter their political identity and influence how they behave in international politics. They presented two coexisting "logics" for how this occurs. First, actors choose a particular action by calculating what consequences are likely to result from that action. Second, actors make decisions by considering who they are (their identity), how they fit within a wider community, and the extent to which their actions are consistent with these identities.

In his classic treatise, Robert Keohane (1984) described how the process might work. He argued that international regimes (networks of international organizations and international treaties that provide rules, norms, and appropriate procedures for member states) helped to regulate the behavior of states by reducing the costs of making and enforcing rules between states as well as by allowing the free flow of information that makes cheating on agreements and deception difficult. Because he was challenging neorealist interpretations of international relations, Keohane's work, and the immense research program it generated, necessarily adopted some of the same assumptions as the neorealists. It is important to note that for neoliberals such as Keohane the focus remained on the state with minimal regard for the role that might be played by individual people.

Jon Pevehouse's (2002, 2005) study of the impact of regional organizations on states' decisions to adopt democratic governance is an example of how this theoretical framework has been used by scholars studying democratic political socialization. Pevehouse (2002) argued that international organizations composed of democratic states would be more effective democratic "socializers" than international organizations that had fewer democratic members.[5] He thought that international organizations that had many democratic member states would be more capable of exerting material and ideational pressure on its authoritarian members than international organizations that had a smaller proportion of democratic members. Using its material resources as well as normative pressure, the international organization could take punitive measures such as suspending trade or diplomatically isolating any offending state. It could also take constructive measures such as reassuring business and military elites that their interests would be protected in any democratic transition (Pevehouse 2002: 522).

In the case of military officers, Pevehouse (2002: 527–28, 536) hypothesized that regional security organizations such as NATO could provide continuing financial benefits and might socialize military officers to an external, rather than internal, focus. Pevehouse's analyses supported his central argument on system-wide effects, but his statistical results did not bear out his hypothesis on military elites; he found that states with a history of rule by military regimes were particularly impervious to the democratizing effects of institutions. Pevehouse's work is important because it shows how both asocial and more "social" explanations of state behavior were emerging within international relations, but once again his analyses were placed primarily at the level of the state.

Another example of research on international institutions and their role in democratic political socialization is the large research program that examines how international human rights institutions regulate states' human rights practices. Emilie Hafner-Burton (2012: 277–80), who has made significant contributions to this literature, found the results of recent scholarship on international human rights regimes both confusing and contradictory. According to her survey of empirical studies, at best, membership in international treaties and organizations on human rights had no impact on state behavior, and, at worst, they correlated with the worst abuses of human rights. She argued that part of the problem with researchers' statistical research designs has been an overreliance on the state as the unit of analysis and research topics that "disproportionately focused away from the places where human rights abuses [were] most severe and institutionalized" (Hafner-Burton 2012: 278). In a poignant criticism of the research program she noted, "Dictators and serial human rights violators are, with few exceptions, outside the mainstream influence of most international human rights legal regimes and thus are increasingly outside the core debates and theories" (Hafner-Burton 2012: 280). She made the important point that researchers have not sufficiently addressed the reality that it is individual people who either abuse or uphold the rights of others:

> Actual people are nearly absent from this research agenda. The exceptions mainly are focused on the judges or commissioners and their decisions, or on voters. Yet institutions by themselves do not abuse human rights; people run state institutions and people abuse human rights. People are important both in terms of how they craft and use human rights legal regimes and in terms of how they decide whether to engage in acts that these regimes prohibit. Neither facet is well understood or currently central to the research agenda in political science. . . . There is tremendous room for research connecting people—especially the people making decisions about whether to engage in human rights abuse—to the incentives and processes playing out in and through international human rights regimes. (Hafner-Burton 2012: 282)

What is particularly interesting about her review of research on international human rights regimes is that Hafner-Burton (2012) does not mention a single study, *not even one study*, that examined how and why state officials, such as soldiers, police officers, or other types of internal security officials, commit abuses. In sum, a key omission in the study of democratic political socialization has been the lack of attention to the beliefs and behavior of actual people.

2.3.2. The Social View of Socialization

The social view of socialization takes into account individuals and how learning, social interactions, persuasive arguments, and normative pressures affect individuals and their communities. In general, the study of how ideas, such as identity, could be used as an instrument of influence in international politics was mostly neglected within international relations until the waning years of the Cold War and the emergence of constructivist scholars who directly challenged the dominant rationalist theories (neorealism and neoliberalism) of state behavior. This post–Cold War trend encompasses an emerging body of literature in international relations that examines political socialization as a process of state identity change (not just behavioral change) with analyses focusing on how state identities are formed and altered through social interactions.

This more social view of socialization that takes into account identities is most closely associated with constructivists, such as Wendt (1999) who described state socialization as a process whereby states came to redefine themselves and others based on their social interactions and their collectively shared ideas about each other and the international system. Teaching and learning were important in these theories. For example, Martha Finnemore (1996: 11) argued in her study of the origins of national interests, "States may not always know what they want and are receptive to teaching about what are appropriate and useful actions to take." In this sense, behavioral changes occur because states learn and adapt and in so doing their identities change and evolve. Wendt's work was both revolutionary and important in that he put forth a more "social theory" of international politics, where states as political entities formed friendships and engaged in rivalries, instead of just selfishly reacting to each other's material capabilities. While Wendt (1999) hypothesized an alternative perspective on the "socialization" of states, he used the state as the unit of analysis,[6] bracketing any role played by individual people.

During the waning years of the Cold War, neoliberal scholars also introduced a more social view of international politics with their discussions of the normative pressures exerted by international institutions. Yet, the extent to which international institutions influenced state behavior independent of the desires of powerful states remained disputed.[7] Scholars hoped to bridge this dispute by incorporating insights from different theoretical perspectives. In this respect, Ikenberry and Kupchan's (1990) work is important because it incorporated both realist material-based arguments as well as constructivist identity-based arguments into a theory of hegemonic socialization.[8] They conceived of state socialization as "a process of learning in which norms and ideals are transmitted from one party to another" (Ikenberry and Kupchan 1990: 289). They argued that policymakers in states targeted by a hegemonic state could be socialized. Policymakers would eventually internalize the norms and ideas transmitted from the hegemonic state, and would then see themselves and their state as a member of an international community in which members shared the norms and beliefs of the hegemonic state. A powerful state could certainly use its military and economic resources to coerce and bribe weaker states to comply with its policies. Yet, they argued, coercion and material incentives were not the only mode or even the most effective way that powerful states used their influence: "There is a more subtle component of hegemonic power, one that works at the level of *substantive beliefs* rather than material payoffs" (Ikenberry and Kupchan 1990: 283, italics in the original).

Ikenberry and Kupchan's concept of normative persuasion emphasizes the role of soft power in international politics. They argued that one way a powerful state could politically socialize potential adversaries was by persuading the elites in the targeted state that other values, beliefs, and norms were superior to their own. A change in norms would subsequently result in policy change in the targeted state making it no longer a threat. Significantly, they highlighted educational exchanges as one effective tool of public diplomacy. Normative persuasion occurred, according to Ikenberry and Kupchan, when a powerful state relied on "ideological persuasion and transnational learning through various forms of direct contact with elites in these states, including contact via diplomatic channels, *cultural exchanges, and foreign study*" (Ikenberry and Kupchan 1990: 290, italics added). While economic and military power remained important, they concluded that there was "a component of power that is not reducible to the coercive capacities of the hegemonic nation. The ability to generate shared beliefs in the acceptability or legitimacy of a particular international order—that is, the ability to forge a consensus among national elites on the normative underpinnings of order—is an important if elusive dimension of hegemonic power" (Ikenberry and Kupchan 1990: 289).

2.4. DEMOCRATIC POLITICAL SOCIALIZATION AFTER THE COLD WAR

After the Cold War, the United States sought to legitimate its place as the world's leading power and to promote democracy as the preferred system of governance. U.S. policymakers needed to put into practice some of the theoretical ideas discussed above. That is to say, they needed to generate shared beliefs in the efficacy of the U.S. vision of international order. As former U.S. President George H. W. Bush wrote in the immediate aftermath of the Cold War, "Let us work together to lead the world toward the 21st Century, the Age of Democratic Peace. There is no more important goal to which we could aspire" (U.S. White House 1993: 21). This goal was also widely shared between European governments that declared in the 1990 Charter of Paris: "We undertake to build, consolidate and strengthen democracy as the only system of government of our nations. . . . Our States will cooperate and support each other with the aim of making democratic gains irreversible" (OSCE 1990: 3–4).

European scholars have been particularly interested in democratic political socialization as it impacted two recent phenomena: the acquisition and support of democratic institutions and practices in Eastern and Central Europe after the end of the Cold War, and the integration of states into the European Union (EU). Compared to U.S. scholars, European international relations scholars have framed more of their research using identity-based arguments rather than rational-instrumental arguments. Important research projects on democratic political socialization include Alexandra Gheciu's study of NATO's democratic socialization of former Soviet bloc countries, Frank Schimmelfennig's research on European regional institutions, and Jeffrey Checkel's work on European integration. All three identify important factors that are incorporated into the arguments presented in this book.

Gheciu (2005) examined how former communist countries in Eastern and Central Europe were integrated into the democratic West through NATO programs that successfully socialized their citizens to democratic norms and practices. Theoretically, she used a constructivist explanatory framework that focused on the mutual constitution of ideas and identity. In her study she conducted interviews with political elites, observed NATO educational programs, and analyzed official publications and documents. She documented how NATO officials combined various ways of inculcating new beliefs, values, and norms in states transitioning from authoritarian to democratic governance. These mechanisms of democratic political socialization included: systematic teaching of elite leaders about democratic governance, educating the public on the value of reform, and persuading leaders to adopt particular democratic reforms. She showed how these efforts were instrumental in the democratic socialization of both elite decision makers and ordinary citizens in Romania and Czechoslovakia.

Education was key to NATO's socializing efforts because former citizens of communist countries had little familiarity or previous experience with how democratic institutions should be organized or how they should function in their lives.

In terms of the socialization theories discussed earlier, they were novices seeking information. For example, in the case of Romania, Gheciu (2005: 986) observed that prior to NATO educational initiatives "there was virtually no understanding of the concept of soldiers as citizens in uniform, entitled to basic rights and bound by a set of duties concerning the protection of the rights of citizens even in times of crisis." Gheciu's study demonstrated the effectiveness of elite education in providing people with the necessary knowledge for building democratic governance and the incentives for adopting it.

Schimmelfennig (2000) took another approach to understanding democratic political socialization. He examined how the EU, NATO, and Council of Europe sought to democratically socialize former communist states through European integration initiatives. The Western community offered international legitimacy and material assistance as incentives for Central and Eastern European leaders who introduced liberal reforms. Schimmelfennig (2000: 117) argued, "Rational political actors regard legitimacy as a resource that strengthens their political power, improves their access to positions of authority, enhances their capacity to govern and extends their period of rule. . . . States calculate whether the benefits of international legitimacy are worth the costs of adaptation and conformity." His study showed that a state might learn and adopt new values and norms "not because it regards them as true and right, but because adoption is necessary to further its political goals." Once adopted, however, these values and norms become ingrained.

Checkel (2001) examined the conditions under which European states chose to adopt or resist EU citizenship concepts being promulgated by the Council of Europe. Checkel's study showed the importance of social interaction and persuasion as mechanisms of influence and he specified the conditions under which these mechanisms functioned most effectively. He argued that persuasion and social learning were key processes leading to norm compliance in addition to rational cost-benefit decision-making. Checkel (2001: 562) defined the term "argumentative persuasion" as a social process of interaction "in which a communicator attempts to induce a change in the belief, attitude, or behavior of another person" through communication and in the absence of coercion. He found that officials in Ukraine, a transitioning democracy with "novice" officials, were receptive to argumentative persuasion, but such was not the case in Germany where policymakers were well conditioned to many appeals and daily pressures and had strong beliefs in their own institutions. This finding again highlights an important factor in socialization: when the target of socialization efforts lacks knowledge or is seeking out additional knowledge, the socialization effort is more likely to be effective.

To briefly summarize, the studies described above identify several factors that facilitate political socialization. First, the education of elite leaders is important because leaders set policies for their own citizens as well as the foreign policies of their countries. Second, efforts to persuade others are likely to be most effective when the person is seeking help or lacks some information. Third, social interaction and social learning are important ways to reinforce persuasive arguments and educational programs. Fourth, while elites may adopt new ideas because they are politically expedient, these new ideas nevertheless become ingrained and accepted over the longer term.

2.5. SOCIALIZATION THROUGH EDUCATIONAL EXCHANGES

Governments have long used educational exchanges as a way to socialize others; and through this process promote their agendas and gain international influence. Educational exchanges continue to be an important tool of public diplomacy, perhaps one of the most influential means of building soft power over the longer term. Exchange participants usually return home with warm memories, new friends, and an increased interest in the places where they lived and visited. They frequently become advocates for their programs and courses of study abroad and remain in contact with their foreign hosts and new friends. They are likely to maintain a lifelong interest in the countries where they lived.

Because educational exchanges are effective tools of public diplomacy, it is not surprising that the United States and the Soviet Union pursued political socialization efforts through educational exchange programs during the Cold War. For example, on the Soviet side, students who studied at the International Lenin School in Moscow were more likely to achieve prominent positions within the communist party in Great Britain during the 1920s through 1940s (Cohen 2005: 229–30). On the U.S. side, U.S.-hosted exchanges exposed future political leaders within the Soviet Union to the freedoms and prosperity found under a democratic government and free market system. Scholarly exchanges increased Western influence particularly within the Russian intelligentsia. The interactions of Soviet students and scholars with Western political, scientific, and academic professionals helped foster gradual liberalization in the Soviet Union that contributed to the eventual collapse of the Soviet system (Richmond 2003; Nye 2004). For example, Oleg Kalugin, who rose to general officer rank in the KGB,[9] spent a year as an exchange student at Columbia University in 1958–1959. Interviewed in 1997, Kalugin noted the importance of such programs in undermining the Soviet system: "Exchanges were a Trojan Horse in the Soviet Union. They played a tremendous role in the erosion of the Soviet system. They opened up a closed society. They greatly influenced younger people who saw the world with more open eyes, and they kept infecting more and more people over the years" (Oleg Kalugin as quoted in Richmond 2003: 32).

In the United States, we tend to think of foreign exchange students as the young people from a wide range of countries who come to study at U.S. universities. Not as visible, however, are the thousands of foreign military personnel who have received elite military education within the United States. The military educational exchanges sponsored by the U.S. military's war and staff colleges are particularly important tools of public diplomacy because their participants are likely to have substantial influence not just within their own circle of friends but within their governments, so the knowledge and memories that they take home could impact both domestic and foreign policies of their countries. Also, military officers frequently control the means that authoritarian governments use to stay in power. Thus, they serve as a critical node in democratic transitions. During the Arab Spring, the Egyptian and Tunisian militaries chose to defend their fellow citizens against the dictatorship, and this led to success for citizens in throwing off the repressive regimes. On the other hand, Syrian military officers chose to support the al-Assad dictatorship, and Syrian citizens along with their democratic aspirations have suffered severely.

Military exchange programs have received little scholarly attention because, as the research reviewed in this chapter shows, scholars have usually viewed military organizations as unitary entities or sub-units within a state. Each state has its own weapons and soldiers that its political leaders use to obtain their political objectives usually through the use or display of its destructive capabilities. In reality, however, military organizations are social structures in which a vast myriad of human interactions take place. Soldiers are people too. They form friendships and interact with the other people in their lives. They exchange ideas about themselves, their work, and their lives. They form social and professional networks particularly with others who share their interests and values. They travel to foreign lands for training and to work with allies. New recruits are trained by their fellow countrymen, but as they move up into the higher ranks, they might also be sent abroad for training. For example, military officers are sent to foreign military schools in order to learn the doctrine and skills that they need to work with allies. Soldiers might attend courses abroad in order to learn how to operate military equipment that their country is purchasing from another country.

For both civilians and military officers from authoritarian countries, the time spent on an exchange program in the United States offers the opportunity to observe how U.S. officers behave within a democratic system, to acquire knowledge about how democratic institutions function in everyday life, to learn about citizens' roles and responsibilities, and to learn what to expect of their own leaders and institutions. These ideas travel across geographic borders when exchange participants return home. Once home, participants may hold their own government institutions accountable through overt actions such as protests. But more likely their influence is very subtle. Some may simply choose not to help repressive leaders gain or stay in power. Others may seek to reform existing practices or political institutions through their own internal policies. Others may already hold politically powerful positions in their governments and can directly alter policies. Others may simply share their experiences with friends and family. As soft power scholar Joseph Nye (2004: 13) argued, these interactions are not trivial: "The ideas and values that America exports in the minds of more than half a million foreign students who study every year in American universities and then return to their home countries . . . tend to reach elites with power."

Educational exchanges are one way that transnational linkages between soldiers are formed. It is through these types of contacts that new ideas are introduced into what is usually thought of as a rather insular organization; in fact, that is exactly the goal of political leaders who send their military officers to study at U.S. military schools. They want their officers to bring new military expertise back home. The curriculum at U.S. military schools focuses on military topics and exposes soldiers to subjects such as U.S. national security strategies, the organization and functioning of U.S. military command and control organizations, and U.S. military doctrine. But there is no reason to assume that when the officers attend class and talk and socialize with their fellow students and instructors, the only ideas they exchange are about the work that they do. They also talk about their everyday lives, their hopes, and their families. Educational exchanges, including military exchanges, provide both the learning and the social environments that Ikenberry and Kupchan (1990), Gheciu (2005), and Checkel (2001) found instrumental in successful democratic political socialization.

2.6. SUMMARY

This chapter discussed how socialization has been studied within the context of international politics and identified key mechanisms that make military educational exchanges an effective environment for political socialization and, hence, for the development of U.S. military soft power. Educational exchanges, particularly military exchanges, are an important asset of U.S. public diplomacy. The military exchanges not only educate and socialize their participants; they also help to build an international professional network of military officers from countries around the world. Political socialization is a key function of educational exchanges in building and sustaining these networks.

NOTES

1. See Niemi and Hepburn (1995) for more details on the demise of the field.
2. The U.S. government's efforts to support democratic governance will be discussed in the next chapter.
3. Thies himself is not a neorealist, but sought to show how Waltz's notion of socialization could be logically consistent with the larger body of neorealist theory.
4. International institutions can be thought of as organizations that states join, such as the United Nations (UN) or the Organization of the Petroleum Exporting Countries (OPEC) and also as agreements between states that set out rules, procedures, and obligations, such as the North American Free Trade Agreement (NAFTA) or the Geneva Conventions.
5. Pevehouse (2002: 536) argued that the greater the percentage of democratic members in an intergovernmental organization, the more likely it would be that a nondemocratic member state of that organization would transition to democracy.

6. In this earlier work, Wendt (1999: 197) states: "I argue that states are real actors to which we can legitimately attribute anthropomorphic qualities like desires, beliefs, and intentionality."

7. See, for example, Mearsheimer (1994–1995), Baldwin (1993), Grieco (1988), and Strange (1982).

8. Ikenberry and Kupchan (1990) evaluated three mechanisms of hegemonic state socialization. First, a powerful state might intervene directly in the internal affairs of another state. Second, the powerful state might use external inducements such as economic incentives or military coercion that would cause a change in policies, rules, or institutions that in time would come to be viewed as legitimate, leading to the acceptance of the norms that underlie them. Third, a powerful state could persuade the elites in a target state that other values, beliefs, and norms were superior to their own. A change in norms would subsequently result in policy change in targeted states.

9. KGB is the English transliteration of the Russian acronym КГБ that stands for Комитет государственной безопасности (Committee for State Security), the Soviet Union's much feared secret police.

3

Building U.S. Military Soft Power

The last chapter examined different theories of how individuals and their communities are socialized. In this chapter these theoretical ideas are applied within the context of military organizations and military power. The purpose is to show the important place that military educational exchanges play in U.S. foreign policy as a source of national power. The chapter begins by describing how the study of international security issues in the United States has been dominated by a conceptualization of military power that emphasizes material capabilities, especially the use of military force to coerce and impose the state's will on others. Theorists and practitioners in this tradition have concentrated their attention on how material resources such as money, weapons, equipment, and personnel can be used most effectively to achieve various political goals from conquest to deterrence. From this theoretical perspective, military organizations are seen as unitary actors with little attention paid to the interactions of the people within these organizations. A brief historical overview of how the traditional view of military power has informed scholarship and U.S. foreign policy illustrates how an over-emphasis on this "traditional view" of military power has been a weakness in U.S. strategic thinking. The chapter then examines a "nontraditional view" of military power that emphasizes the impact of shared ideas, collective identity, and socialization.[1] This perspective on military power has received more attention in recent years because the United States' involvement in wars in Iraq and Afghanistan called for greater attention to how ideas, such as culture and identity, can be used as a power resource.

The chapter contends that the U.S. military exerts influence as part of an international military community in which military exchanges play an important role. The central claim is that in addition to their function of fighting wars, deterring conflict, and coercing adversaries, the U.S. military has built and nurtured an epistemic community through which ideas, identities, norms, and beliefs are exchanged and learned. The chapter discusses a variety of forums, in addition to military educational exchanges, where soldiers from different countries interact, form friendships, social networks, and the larger epistemic community. For U.S. foreign policy this military community is important because it helps to build goodwill toward the United States, creating U.S. soft power.

3.1. THE TRADITIONAL VIEW OF MILITARY ORGANIZATIONS

Military organizations are typically viewed as agents or instruments of states, and their influence is estimated by counting how many weapons, soldiers, and applications of the latest technology are incorporated within them. Robert Gilpin (1981: 13) put it most succinctly: "Power refers simply to the military, economic, and technological capabilities of states." This traditional view of military power has long dominated the study of how military organizations influence international politics. It is evident in U.S. government strategic doctrine such as the 1997 National Military Strategy of the United States, which portrays the military as an instrument of the state used to deter adversaries and fight and win the nation's wars:

> The Armed Forces are the Nation's military instrument for ensuring our security. Accordingly, the primary purpose of U.S. Armed Forces is to deter threats of organized violence against the United States and its interests, and to defeat such threats should deterrence fail. The military is a complementary element of national power that stands with the other instruments wielded by our government. The Armed Forces' core competence is the ability to apply decisive military power to deter or defeat aggression and achieve our national security objectives. (U.S. Chairman of the Joint Chiefs of Staff 1997: introduction)

In the traditional view, the military is seen as a unitary actor that is used as an instrument of power by one country to impose its will on another country. Military power is defined in terms of military capabilities from weapons to monetary resources to strength of alliances. In order to assess potential military power between different countries, traditionally oriented analysts compare military equipment, number of soldiers, other material assets, and how allies might augment such capabilities. Within the study of international relations, such assessments of military power can be seen in the publication of annual assessments of the military capabilities of different countries such as the *Correlates of War Project* founded by J. David Singer at the University of Michigan, the Stockholm International Peace Research Institute's annual count of worldwide armaments subtitled *Armaments, Disarmament and International Security*, the Peace Research Institute Oslo's Centre for the Study of Civil War's armed conflict database *Armed Conflicts*, and the International Institute for Strategic Studies' annual publication *The Military Balance*. All of these data sources focus on counting material resources such as manpower, equipment, defense spending, and the like, in order to quantify military power.

In international relations scholarship, this conceptualization of military power particularly resonated during the Cold War. Neorealist scholars such as Waltz (1979: 98) believed that "power is estimated by comparing capabilities." Emphasis on relative capabilities prevailed in U.S. government analyses, such as this recently declassified U.S. government assessment of the Soviet Union published during the height of the Cold War by the Central Intelligence Agency:

> For more than two decades, the USSR has engaged in a major buildup of its military forces. In the Khrushchev era the emphasis was on strategic nuclear programs, but since Brezhnev came to power in 1964 there had been an across-the-board expansion and modernization of all the Soviet forces. . . . Annual Soviet military spending has nearly doubled in real terms and now consumes over one-eighth of GNP; military manpower has increased by one-third to more than 5 million; defense research and development facilities have more than doubled in size; and weapon production facilities have expanded by nearly 60 percent. The number of Soviet strategic nuclear weapons delivery vehicles has increased from a few hundred in 1965 to about 2,500 today, overturning the previous U.S. quantitative superiority. (The United States has just over 2,000 delivery vehicles.) The accuracy of the newest Soviet weapons now exceeds that of U.S. systems, creating a major threat to U.S. fixed, land-based missiles. (U.S. Central Intelligence Agency 1981: xiii)

The idea that the military is a collection of material resources that is used as an instrument of power has deep historical roots. Perhaps the most influential (and most quoted) is the great military theorist Carl von Clausewitz (1832: 87) who put forth as a central thesis the idea that "war is not merely an act of policy but a true political instrument, a continuation of political intercourse, carried on with other means." In the early twentieth century this view gained increased prominence in Western Europe and the United States with the failure of liberal ideals, embodied in the demise of the League of Nations, as well as with the industrialization of armies. The mass production of tanks, airplanes, aircraft carriers, submarines as well as technological innovations in command systems provided a greatly improved ability to control and direct increasingly lethal military forces across vast distances. Long-range missiles entered military arsenals; and, in conjunction with the acquisition of the thermonuclear weapon in 1952, gave the United States, and soon its chief rival, the Soviet Union, the ability to reach far beyond their geographic borders with tremendous destructive capability. Nuclear weapons were essentially too destructive to use, so civilian strategists argued that such weapons should be used short of war to threaten, bluff, negotiate, and deter. To make such arguments, strategists developed theories that relied on assumptions of rationality, instrumental cost-risk calculations, and assessments of relative material power. Most famously, Thomas Schelling (1960 and 1966) used rational value-maximizing ideas from game theory to show how nuclear weapons could be used effectively, not on the battlefield, but as a bargaining resource.

During the Cold War, the United States based its national security strategies first on arguments for superior quantity of the most destructive forces, which evolved to arguments about the superior technological quality of Western military forces, and finally to arguments on the ability to devote significantly more resources to developing new weapons and new defenses in space. Comparisons were made in the quantities and qualities of these material resources, in the skill of military commanders in employing their forces, as well as in the rationality of political leaders to use the military capabilities; yet, within all these arguments, the focus remained on military power defined in terms of material resources. This is not to say that international relations scholarship completely ignored other views that did not fit within this paradigm, but only to note that attention to factors such as human interactions at the individual and community levels was minimal and certainly not within mainstream international relations.[2]

The quintessential portrayal of the traditional view of military power was the characterization of the United States and the Soviet Union as superpowers racing to build more and more armaments. In his memoirs, renowned Soviet analyst of U.S. foreign policy Georgi Arbatov[3] recalled the dynamics of the resultant security competition:

> [T]he 1970s marked the Soviet Union's most intensive postwar efforts to develop and accumulate arms of all types . . . we reached absurdly high ceilings in many areas. This applies to such important components of conventional forces as the number of people under arms; tanks; artillery; tactical missiles; many types of aircraft; submarines; and many other weapon systems. As far as nuclear arms are concerned, we surpassed the Americans in the number of delivery systems, megatonnage, and throw weight in strategic arms, and also in medium range weapons. The Americans acted equally irresponsibly with regard to some other weapons. It really often looked like a competition in absurdity. (Arbatov 1992: 203)

This traditional view of military power based on comparisons of the number, range, and firepower of strategic nuclear weapons as well as the number of troops stationed in the European theater dominated foreign policy discourse in the United States during the Cold War. For example, U.S. President Ronald Reagan characterized the Soviet Union's "unprecedented military buildup" of material assets to include "uniformed military of more than five million" and military expenditures totaling "15–17 percent of the total Soviet GNP" as the most significant threat to U.S. security (U.S. White House 1987: 6).

Within the discipline of international relations, scholars characterized international conflict as the result of one country maneuvering to build their own power and balance the power of their rivals. Within this paradigm, both realist and neoliberal scholars understood power to be measured in terms of material capabilities of which military capabilities were the most important. For example, Robert Keohane (1984: 40) argued, "The military conditions for economic hegemony are met if the economically preponderant country has sufficient military capabilities to prevent incursions by others that would deny it access to major areas of its economic activity." Policymakers and military leaders believed that the United States would prevail in any military confrontation in which it had the overwhelming preponderance of military capabilities. Conflicts in Korea, Vietnam, and later in the Middle East challenged this traditional view of military power.

In the Korean War, a small country, the Democratic People's Republic of Korea, fought the United States and its nuclear arsenal to stalemate, but Chinese involvement in the conflict and the threat of Soviet involvement resulted in the traditional U.S. view of military power remaining intact. As Korean War historian Max Hastings (1987: 333–34) wrote, "It may be argued that the United States learned many of the wrong lessons from the experience of Korea, where a huge commitment of resources enabled her army to halt the Communists at tolerable cost in casualties. . . . The American Army emerged from Korea convinced that its vastly superior firepower and equipment could always defeat a poorly equipped Asian army if it was provided the opportunity to deploy them." From the perspective of the United States, the Korean War simply reinforced the idea that material resources were key to victory; when victory appeared elusive more military resources were needed. Military historian Sir Michael Howard (2004: 124) noted that the Korean War provided the impetus to remilitarize (after post–World War II decline) the United States and reinforced traditional views of military power: "up till June 1950 there was little sign that the American people, whatever their noble intentions, were prepared to pay the price of being a world power. Korea changed all that. Defence appropriations were quadrupled almost overnight and continued to soar."

A decade later, the Vietnam War once again challenged the traditional U.S. view of military power. Only this time, the adversary's skillful use of propaganda in addition to asymmetric strategies undermined U.S. public support for the war and brought starkly into public view how ideas such as justice, legitimacy, and morality might trump material resources. In the United States, the expectation had been that superior technology and superior firepower should and would prevail against weaker adversaries. Yet, neither quantity of weapons nor technological superiority allowed the United States to achieve its political objectives through the use of armed force in Southeast Asia. A few years after U.S. involvement in the Vietnam War ended, military historian Russell Weigley (1977: 477) concluded his classic treatise on the American way of war with the prescient statement: "Because the record of non-nuclear limited war in obtaining acceptable decisions at tolerable cost is also scarcely heartening, the history of usable combat may at last be reaching its end." Nina Tannenwald (2008: 240) articulated the sentiment for many political scientists: "One of the major lessons of Vietnam for students and practitioners of international relations has been the normative and political limits on material power."

In the years following the Vietnam War the use of military force by the United States became increasingly proscribed for a number of reasons. Richard Rosecrance (1986) reasoned that because countries valued economics as the most important area of competition, military force was less useful. Democratic peace scholars led by Michael Doyle (1986) pointed to the spread of democratic values and governance as a restraint on warfare, at least within the community of democratic countries. Also important was the growth of international institutions that helped countries resolve their disputes short of war. John Mueller (2004) predicted that traditional international war would continue to decline as developed nations chose to resolve their conflicts in other ways. Mueller (2004: 161) pointed out, attitudes toward war had changed "as developed countries seize control of their destinies and decide that war with each other should not be part of them."

In this broader context, the U.S. military faced a difficult dilemma: as the political acceptability of the use of large-scale military battles as an instrument of national power declined, the U.S. military was nevertheless tasked with ensuring the security of the United States, from shore to shore and overseas. U.S. military leaders still remembered their initial near defeat in Korea when material capabilities of all sorts were in short supply following demobilization after World War II, and in Vietnam where political objectives prevented the use of maximum military force. In the aftermath of U.S. withdrawal from Vietnam, new theories (and rationalizations) were needed to explain why a military superpower could not defeat an economically underdeveloped and militarily inferior adversary. For its own part, the U.S. military considered its defeat in the Vietnam War an anomaly and continued to emphasize overwhelming military force as decisive to victory in war. The U.S. Army, reflecting the Cold War goals of its political leaders, continued to worry about large-scale land war in Europe: "Battle in Central Europe against forces of the Warsaw Pact is the most demanding mission the U.S. Army could be assigned. . . . The U.S. Army is structured primarily for that contingency and has large forces deployed in that area" (U.S. Department of the Army 1976: 1–2). The U.S. Army focused on fighting major conflicts and adopted the concept of "AirLand Battle," which emphasized material resources: combined arms and working jointly with the Air Force, Marine Corps, and Navy. Such strategic thinking was vindicated and reinforced by the highly effective use of U.S. military forces in operations against Iraq in 1990–1991. The one-hundred-hour operation to expel Iraqi forces from Kuwait focused on military power as traditionally conceived: the deployment and meeting of military equipment, weapons, and soldiers on a battlefield. In Operation Desert Storm, the easy triumph of superior military capabilities against a more poorly equipped adversary seemed to vindicate those who had argued that Vietnam was an anomaly. Superior material resources, in the end, were victorious; it was only a matter of retaining support of the American people and the ability to use overwhelming force to encircle and annihilate the enemy.

The vindication, however, was short lived. The 2001 war in Afghanistan that lingers on, and the 2003–2011 war in Iraq did not follow the pattern of the 1990–1991 Iraq war. These wars raised questions about the over-emphasis placed on traditional concepts of warfare to the near exclusion of considerations of culture, identity, norms, and other nonmaterial factors. As U.S. politicians debated and made the decision to conduct military operations against Iraq, they believed that U.S. military capabilities in terms of firepower and advanced technologies would be able to sweep away vastly inferior Iraqi forces, with victory ensuing in a matter of weeks, if not days. In less than two months after combat operations began, U.S. President George

W. Bush declared "mission accomplished" from the deck of the aircraft carrier *USS Abraham Lincoln*, a quintessential symbol of the traditional view of U.S. military power. Sitting atop this enormous arsenal of material capabilities, the president declared victory; but victory conceived in terms of traditional ideas of military power. While the United States reigned materially supreme, it was soon losing an ideational battle that in the following years threatened to nullify its material victory. Instead of winning the war on the basis of superior manpower, weapons, communication systems, and information technologies, the United States struggled with how best to wage a "war of ideas." Political leaders and military commanders sought strategies that might help them translate material superiority into viable ideas, such as legitimacy and justice, that resonated and motivated. The so-called "war of ideas" for "hearts and minds" was to prove troublesome and difficult. One member of a White House public diplomacy advisory group commented to *The Washington Post*, "It's worse than failing. Failing means you tried and didn't get better. But at this point, three years after September 11, you can say there wasn't even much of an attempt, and today Arab and Muslim attitudes toward the U.S. and the degree of distrust in the U.S. are far worse than they were three years ago" (Wright 2004: A01).

3.2. RENEWED FOCUS ON THE IDEATIONAL ASPECTS OF MILITARY POWER

Recent military conflicts that were less amenable to resolution through only traditional military means resulted in renewed interest in nonmaterial aspects of power. The poignant anecdote told by soldier-turned-scholar Harry Summers about the Vietnam War resonated with military commanders in the early twenty-first century. As Summers related the story; when he returned to Vietnam in 1975 as part of a U.S. negotiating team, he told the head of the North Vietnamese Delegation Colonel Tu: "You know you never defeated us on the battlefield." The North Vietnamese Colonel responded, "That may be so, but it is also irrelevant" (Summers 1982: 7).

Summer's comment illustrates the emphasis placed on victory in terms of the clash of military forces on the battlefield by the United States, and betrays the inability to think in larger strategic terms. Although it is true that the United States military was never defeated on the battlefields of Southeast Asia, in the end, all of the United States' battlefield victories were irrelevant because the United States was defeated strategically, that is to say in the minds of U.S. citizens and policymakers. As the great Chinese strategist Sun Tzu might have commented: the most skillful general does not waste his resources in defeating the enemy in warfare, but uses his resources to psychologically outmaneuver the enemy to achieve his goals without resorting to battle.

In the traditional view of military power victory in war should go to the army that retains the material resources and will to continue the fight. However, scholars such as Johnson and Tierney (2006) have shown that ideas of what constitutes victory may be just as important as material aspects in understanding victory. In their work they showed how preconceived ideas about what victory should be like determines perceptions of victory and defeat in warfare, rather than the physical facts on the ground such as the destruction of an enemy's army or cities. In this sense, power is not always evident in observable material capabilities, but in understandings socially constructed through public discourse.

Soldiers and politicians have not been alone in placing emphasis on power conceived of in terms of material capabilities. Scholars, such as George Lawson (2008: 19), have criticized the discipline of international relations as entrenched in old theoretical paradigms that do not "speak to the world within which [international relations] is situated and which it is charged to study." Michael Barnett and Raymond Duvall (2005: 39–40) lamented that post–September 11 debates within international relations were conceptually limited to "a realist conception of power—the ability of states to use material resources to get others to do what they otherwise would not." Richard Ned Lebow (2009: 2) argued, "Any theory of international relations must build upon or be rooted in a theory of society and must address the constitution of actors, not only their behavior." One of the problems Lawson (2008: 18) noted is the inadequacy of existing work on the ideological aspects of power: "It is troubling quite how incapable much of the discipline has been at speaking to the imminent range of puzzles, questions and problems posed by the hows, whys, and wherefores of the contemporary conjuncture: the extent to which the essential grammar of world politics has been disturbed by the shift in the political imaginary from Westphalia to multiple, often deterritorialized, political spaces, [. . .] and from a discussion of material capabilities to relative intangibles" such as religion, race, and culture.

Joseph Nye's concept of "soft power," or the ability to persuade and co-opt, is the most well-known alternative theory of power within international relations. Nye (2004: 5) argued that people (and their countries) are not motivated just by threats and bribes but also by ideals: "A country may obtain the outcomes it wants in world politics because other countries—admiring its values, emulating its example, aspiring to its level of prosperity and openness—want to follow it." In the longer term, these motivations can be more enduring because they change identities, beliefs, and the fundamental ways in which people see themselves and their world. Perhaps this was in the mind of Fidel Castro when he remarked as he was retiring from political leadership: "I am not saying goodbye to you, I only wish to fight as a soldier of ideas" (McKinley 2008).

Castro's remark reflected the growing emphasis that political leaders attributed to how ideas exerted influence, not only at the periphery of politics but at its core, in how we think about military power itself. Scholars have shown that collective beliefs and notions about morality impact political decisions and political behavior, and even decisions at the heart of national security such as how to employ nuclear weapons. As Tannenwald (2008) showed, ideas about appropriate behavior, legitimacy, and morality have helped to prevent the use of nuclear weapons since World War II.

The Cold War and its nuclear and conventional arms races certainly emphasized power in terms of material resources. Yet, to say this was the exclusive focus of policymakers would not be totally accurate. The Soviet Union's interactions and engagement around the world concerned U.S. leaders. President Ronald Reagan may have identified the Soviet buildup of military capabilities as the most significant threat, but he also worried about other aspects of Soviet power; in particular he worried about the Soviets' "massive political influence apparatus" that included the "political and ideological indoctrination of foreign students, government officials, terrorists, and military personnel; and perceptions management of foreign visitors to the USSR" (U.S. White House 1987: 7).

More recently, the George W. Bush Administration might be best remembered for its proclivity to use unilateral military force in the wake of the 2001 terrorist attacks, yet its national security strategies also spoke of the ideological struggle taking place between the United States and its terrorist adversaries. In its 2002 National Security Strategy, the George W. Bush White House declared that it would "wage a war of ideas to win the battle against international terrorism." The war of ideas included working with allies to delegitimize terrorism and "using effective public diplomacy to promote the free flow of information and ideas to kindle the hopes and aspirations of freedom" in order to undercut the bases of terrorist support (U.S. White House 2002: 6). Four years later the George W. Bush White House again emphasized in its national security strategy that "[f]rom the beginning, the War on Terror has been both a battle of arms and a battle of ideas" (U.S. White House 2006: 9). Yet, his administration failed early on to effectively wage both the battle of arms and the battle of ideas. The inadequacy of focusing nearly exclusively on the battle of arms was highlighted by the U.S. invasion and ongoing war in Iraq as well as the United States' attempts to conduct a "war on terror." Initially, the strategies used in these wars emphasized the use of overwhelming force. Pulitzer Prize–winning foreign policy analyst Thomas Ricks spent months embedded with American forces in Iraq and chronicled just how difficult it seemed for U.S. military and political leaders to adopt new strategies to meet these new threats. He wrote,

> Pentagon officials liked to talk about "rapid decisive operations." . . . The technocentric notion behind it was that U.S. forces, taking advantage of advances in sensors, communications, computer technology, and long-range weaponry and precision logistics, all areas in which it excelled, would fight so quickly and adeptly that the enemy would never have a chance to catch up and understand what was happening. Blinded, confused, and overwhelmed, the enemy's will would break, U.S. forces would triumph, and everyone would live happily ever after. [. . .] [Counterinsurgency expert] Col. McMaster would later argue that the very concept of rapid decisive operations had hamstrung American commanders as they entered the country, because it had "artificially divorced the war from its political, human, and psychological dimensions." [. . .] After four years of failure in Iraq, the U.S. military began to find effectiveness—at least tactically—as its leaders finally became resigned to the reality that *the way to*

> *success was conducting slow, ambiguous operations that were built not around technology but around human interactions.* (Ricks 2009: 160–62, italics added)

Military and political leaders struggled with how to handle this changing nature of warfare. In 2004, *The 9-11 Commission Report* (U.S. Government 2004: 375) devoted an entire section to recommendations for how the United States should "engage the struggle of ideas." Yet, after five years of fighting in Iraq, high-level U.S. officials such as the Vice Chairman of the U.S. Joint Chiefs of Staff acknowledged that "ideas can't be eliminated by simply using guns" and that "we still haven't come up with a good enough strategy or even enough good ideas for competing in the war of ideas" (Gilmore 2006). Former Secretary of Defense Donald Rumsfeld (2006), speaking at the U.S. Army War College, also echoed this negative assessment: "If I were grading I would say we probably deserve a 'D' or a 'D-plus' as a country as to how well we're doing in the battle of ideas that's taking place in the world today."

In the wake of such criticisms, U.S. efforts turned to greater emphasis on culture, values, and beliefs. The strategy eventually adopted in Iraq, known as "The Surge" for the increase in troop levels that it required, called for troops to remain in the towns and villages after they had been secured from insurgent forces. A good portion of the task relegated to U.S. soldiers was to develop contacts and social networks among the townspeople so that they could learn to identify friend from foe. The key was building a strong sense of security among the population so ordinary people would feel safe in providing information on insurgent activities. U.S. military soldiers needed to work with local leaders, the Iraqi military, and townspeople to ensure that areas cleared of insurgents would remain secure. Human interactions between U.S. soldiers and Iraqis, as Ricks said, were key to the strategy.

3.3. MILITARY EXCHANGES AS A SOFT POWER RESOURCE

Human interactions on a person-to-person basis can indeed have important effects on how people think and behave; soldiers are no exception. Clausewitz's portrayal of war as an instrument of statecraft receives a lot of attention, but what is less well known is that Clausewitz was also a proponent of developing military-to-military ties. He recommended actively seeking out military officers who had combat experience and learning from them. While these officers might be in one's own military, this might not necessarily be the case. The expertise might be in other countries that had been engaged in recent wars. In *On War*, Clausewitz advocated developing contacts with experienced officers in other countries and sending one's own officers to observe how other military organizations operated. He argued that these military-to-military interactions would be a useful way for military officers to gain expertise in military strategy and to increase their understanding of the intricacies of the battlefield. The interactions would have a positive personal and professional impact not only on the officer who participated, but also on the officer's peers and subordinates when he shared his new knowledge and ideas with them. Clausewitz wrote,

> Another very useful, though more limited, way of gaining familiarity with war in peacetime is to attract foreign officers who have seen active service. Peace does not often reign everywhere in Europe, and never throughout the whole world. A state that has been at peace for many years should try to attract some experienced officers—only those, of course, who have distinguished themselves. Alternatively, some of its own officers should be sent to observe operations, and learn what war is like. However few such officers may be in proportion to an army, *their influence can be very real. Their experience, their insights, and the maturity of their character will affect their subordinates and brother officers.* Even when they cannot be given high command they should be considered as guides who know the country and can be consulted in specific eventualities. (Clausewitz 1832: 122–23, italics added)

As Clausewitz argued, participating in military-to-military interactions is an important way for military officers to learn new ideas and share them across borders. As he says, the exchange officers might have substantial influence on their fellow officers and subordinates because of the ideas that they bring with them from their time abroad. Thus, even in his treatise that emphasized the instrumental use of military forces, Clausewitz recognized the importance of the military community and military exchanges as channels of ideas that transcend borders and could affect wars' outcomes. Clausewitz's insight was not just a theoretical proposition that he put forth, but was based on his own personal experience. After the defeat of the Prussian army by Napoleon and Prussia's alliance with France, Clausewitz resigned his commission in the Prussian army and volunteered for service in the Russian military, lending his expertise to Russia in the fight against Napoleonic France.

Countries have participated in military exchanges throughout history, as Clausewitz reminds us, but for the United States the exchanges became more prominent with the end of the Cold War. The United States, alone and through alliances such as NATO, has sought to link military organizations and soldiers across geographic borders. Higher ranking U.S. military officers have nearly all had experience working alongside foreign military officers because the U.S. military's professional education system is structured to provide these opportunities particularly for those officers likely to be promoted to high command or otherwise hold influential positions. The U.S. military has developed doctrine, strategies, and a variety of programs aimed at building ties with foreign militaries and their officers. Today, military officers from different countries routinely meet not only through exchange programs but also in many different forums, including participating in peacekeeping operations, working together in common alliances, meeting at professional conferences, participating in training exercises, and attending military schools (Reveron 2010).

To give a few examples, U.S. soldiers have had the opportunity to work with soldiers from other countries within longstanding alliances such as the North Atlantic

Treaty Organization (NATO) and Combined Forces Command (CFC) in Korea. Apart from U.S. membership in multinational organizations, U.S. soldiers also spend a good portion of their career stationed or deployed abroad as part of U.S. military units, where they interact with host country soldiers. U.S. military officers who are in logistics and materiel acquisition fields develop contacts as they negotiate contracts and help other countries purchase and maintain U.S. military equipment.

All of these different situations provide opportunities for U.S. soldiers to interact with their counterparts in other countries. These interactions are the embodiment of key socialization mechanisms discussed in the last chapter: education, learning, social interaction, social networking, and persuasion. Military knowledge, doctrine, and tactics are discussed in these forums, but these forums are also opportunities for the soldiers to build personal friendships and professional networks. As they do so, these military-to-military networks are the foundation of an international military community that can have important socializing impacts.

3.3.1. Epistemic Communities and Their Influence

For soldiers, most of their personal and professional interactions are likely to take place between compatriots. Those are the people they work with on a daily basis; those are the people who they are likely to count as friends when off-duty and with whom they will go to war. However for soldiers stationed overseas or for those who work in multinational organizations, these same types of interactions take place between soldiers of different countries. Friendships and professional networks are formed between soldiers of different countries for much the same reasons as those between compatriots: common interests, similar lifestyles, and the mutual support provided by friendships within a larger community. Peter Haas (1992: 3) called these types of professional communities "epistemic communities," which he defined as a "network of professionals with recognized expertise and competence in a particular domain and an authoritative claim to policy-relevant knowledge within that domain." The members of epistemic communities are linked together through common experiences and shared expertise. As Haas and other scholars have shown, epistemic communities can have significant influence on the decisions made by political leaders because the people in these networks are considered experts in their field and have contacts in important decision-making organizations. Epistemic communities, according to Haas, are repositories of specialized knowledge that state leaders may draw upon in order to identify salient issues, define national interests, and formulate policies. The communities also have an important role in socializing members and aspiring members. Members within an epistemic community share specialized knowledge and develop a common understanding of how to approach and solve problems. They speak in a commonly shared yet specialized language, and within their subject of expertise share common values and priorities (Haas 1992: 3). As new members join the community they are socialized to its norms, values, and ideas.

The study of epistemic communities has mostly focused on the role played by scientists and how they use their expertise to influence policy decisions. For example, Emanuel Adler (1992) showed how, in the 1950s and 1960s, an epistemic community of civilian strategists, engineers, physicists, and other types of scientists were able to transform U.S. and Soviet strategic thinking to the point that it eventually led to the signing of the 1972 Anti-Ballistic Missile (ABM) Treaty between the United States and the Soviet Union and the "stabilization" of nuclear deterrence between the two superpowers. Adler's study showed how individuals and their communities were mutually socialized. His study demonstrates how many of the socialization mechanisms discussed in the last chapter work in real life.

According to Adler (1992), the nuclear arms control epistemic community began as various disparate groups of civilian strategists and scientists who believed that the basis of U.S. nuclear strategy should be nuclear arms control rather than nuclear superiority or victory in nuclear warfare. While the community's members came from different backgrounds and had intellectual disputes and personal conflicts, they nevertheless frequently met and exchanged ideas. Through these intellectual exchanges, they refined their ideas and came to share a common belief that strategic nuclear arms control was essential for U.S. national security. This epistemic community then used their influence within the highest ranks of the U.S. government to change how policymakers thought about nuclear war. According to Adler (1992: 102), "These strategists and scientists reached into the places where decisions [were] made and into the minds of the people who made them, thereby turning their ideas into widespread national security policy and practice. They also were instrumental in diffusing this understanding to the Soviet Union." The epistemic community was able to influence policy because many of its members were brought into the U.S. government where they were able to persuade others of their ideas and institutionalize these ideas into government policies (Adler 1992: 142). As Adler noted, once set in U.S. policymakers' minds, the epistemic community then helped to spread their belief in the efficacy of strategic nuclear arms control to the Soviet Union through various international forums such as negotiations, technical conferences, scientific meetings, and summits; as well as through U.S. national discourses such as press reporting, government statements, and academic writing (Adler 1992: 133–40). This international diffusion began in the 1950s and took over two decades (using the signing of the 1972 ABM Treaty as indicator of the successful diffusion of the ideas to the Soviet Union).

Much like scientists, military officers have specialized knowledge. They are experts on weapons capabilities, military tactics, military strategies, and military doctrines. They often have considerable insight and access into the governing institutions of their countries. Also like other professionals, military professionals interact across borders, forming international networks through which ideas are exchanged, new ways of doing tasks are learned, and normative standards influence behavior. Within alliances and other forums where soldiers from different countries work together, military officers draw on each other's expert knowledge. They may also develop shared values and beliefs within the military community that transcend the borders of their individual countries. In fact, the purpose of alliances and coalitions are to do just that: to develop shared doctrines, strategies, and organizations in order that the militaries of different countries might operate effectively together in wartime and during other military operations.

Haas (1992: 16) argued that epistemic communities, or networked communities of like-minded professionals, "may succeed in imposing their views and moving [decision makers] toward goals other than those initially envisioned by the decision makers." Adler's (1992) study of the development of nuclear arms control showed just how influential epistemic communities might be even in an area of vital national security concern between rival countries.

3.3.2. The Military as an Epistemic Community

Military-to-military interactions can augment military power, as traditionally conceived, by socializing participants and building epistemic communities of like-minded military professionals. There are a number of ways that soldiers come into contact with each other, often for extended periods of time, that help to build a military epistemic community. Alliances are important venues where professional contacts between U.S. military officers and their international counterparts are built. These organizations improve the effectiveness of multinational operations by developing common doctrine and establishing common operating procedures between militaries. In turn, these commonly held ideas (doctrine, strategy, operating procedures) impact how individual countries use their military forces. For example, for over half a century the U.S. and South Korean militaries have worked side by side in Combined Forces Command (CFC) in the Republic of Korea (ROK). CFC has a deeply integrated command structure tasked with the development of warfighting plans and operational doctrine as well as the day-to-day maintenance of warfighting capabilities. For soldiers on a personal level, this means that U.S. and ROK soldiers share common experiences and build common understandings through working closely together day after day. They share offices, often sitting in desks next to one another. They practice their warfighting skills together in combined operational exercises. When the need arises, they participate in military operations together. Sometimes they develop personal friendships that extend beyond the work environment to spending off-duty time together. Thus, the formal work environment may result in personal friendships that transcend the military assignment where they work together.

The North Atlantic Treaty Organization (NATO) is another example of a long-standing alliance in which military personnel from different nations have worked closely with each other for decades. During the Cold War, NATO had a number of programs and organizations in which military officers from NATO member countries met and worked with their counterparts. Close personal and professional interactions between U.S. military officers and their NATO counterparts were beneficial, even necessary, in building and maintaining the alliance. Generations of NATO military officers from many different countries literally spent their careers working together to build NATO institutions, organizations, doctrines, plans, and command structures. Today NATO has been extended to include many diverse nations under programs such as the Partnership for Peace that engages militaries from former Warsaw Pact countries, and the Istanbul Cooperative Initiative that seeks to build "a network of partnerships" initially open to the six countries of the Gulf Cooperation Council but now open to all countries in the Middle East.[4]

Coalitions are another way in which militaries from diverse types of countries work together and build communities of like-minded professionals. In Operation Desert Storm, the United States worked hard to form a coalition that, by necessity, consisted of authoritarian countries such as Saudi Arabia and Kuwait as well as liberal countries such as the United Kingdom and Canada.[5] Since 2001, the United States has expanded its cooperative efforts with military forces from a wide array of authoritarian countries such as Kazakhstan and Pakistan, as well as Qatar and Afghanistan.

In addition to formalized alliances and less formalized coalitions, military officers may also meet in a variety of other venues such as in the bilateral administration of financial aid programs, the negotiations for the purchase of weapon systems, the basing of U.S. troops in other countries, and educational exchange programs hosted by U.S. military professional schools. Each of these types of military-to-military interactions has the potential to socialize participants and incorporate them into a wider social and professional network. One example of the influence that can be brought to bear through this community is in the deployment of U.S. forces abroad. For example, the U.S. Department of Defense (1995) argued that such military ties were helpful in the 1994 deployment of U.S. military troops to the Persian Gulf: "The close military-to-military relationships built up over many years with each of the GCC [Gulf Cooperation Council] states created the environment that allowed host countries to accept the United States' crisis deployment promptly and support it effectively."

Military epistemic communities have also made it easier to build effective coalitions between countries that might otherwise have had little or no operational experience together, such as the building of the military coalition during the Desert Shield/Storm operations in the Persian Gulf region in 1990. A study by the U.S. National Defense University attributed the extensive cooperation between U.S. military officers and their Arab counterparts to the personal relationships and professional expertise learned when these officers had attended training courses together in the United States (Cope 1995: 30). Military epistemic communities have also been shown to increase effectiveness in nontraditional military operations, such as helping to provide humanitarian assistance in the wake of natural disasters. For example, previously established relationships between military officers were instrumental in the quick humanitarian response to assist Kurdish refugees in 1991. Interviews and surveys of participants by the U.S. National Defense University showed that "both the State and Defense Departments benefited from close relationships among several

US foreign service officers, US military officers, and their Turkish military counterparts, all of whom had met as students at US war colleges" (Cope 1995: 28).

Military personnel from around the world regularly attend a number of education and training institutions throughout the United States. Training programs may range from short courses such as how to repair electronic equipment to longer courses such as pilot training. While there are many such schools that provide technical training and short courses, the U.S. military also hosts a large number of foreign military officers as students in its professional military schools: the war and staff colleges.

The war and staff colleges are the U.S. military's elite institutions. For both U.S. officers and their foreign counterparts, attendance is for an entire year, and many officers bring their families with them. Many officers live in local U.S. communities rather than on the military base. In considering the scope of foreign participation at U.S. elite military schools, the National Defense University's Institute for National Strategic Studies noted, "There is no comparable historical example of so many diverse sovereign states augmenting the professional development of their armed forces by entrusting so many potential national leaders to the education and training of another state" (Cope, 1995: 12).

The military epistemic community evolves and grows as new generations of military officers build personal and professional social networks with higher-ranking officers in their own country and with their counterparts from other countries. Programs such as the educational exchanges at U.S. military war and staff colleges are particularly important because they build trust, common knowledge, and working relationships, and serve as socializing environments in which ideas, values, and beliefs might be exchanged both officially and informally. Within these programs military officers from around the world interact, many forming friendships that last a lifetime.

Nye and Owens (1996: 31) argued that government programs such as U.S.-hosted military education and training serve as an influential and cost-effective way for the United States to employ soft power, particularly to help countries that have made the transition away from authoritarianism consolidate their new democracies. These exchanges help to build a wider epistemic community and serve as socializing environments in which ideas about democratic rule and civilian control of the military can reach participants from nondemocratic countries.

3.4. BUILDING MILITARY SOFT POWER

From the perspective of the United States, military educational exchanges have at least three important impacts on U.S. foreign policy: as a source of information, as a source of soft power, and as a source of support for democratic reform. First, the exchanges and the resulting professional and personal networks can serve as an important source of information. Like other sorts of foreign exchange programs, the exchanges hosted by U.S. military schools can help to build goodwill and understanding between the host country and the exchange participants because those involved learn more about each other's backgrounds, motivations, and goals. In its 2010 *Quadrennial Defense Review Report* the U.S. Department of Defense emphasized the need for the U.S. military to foster engagement with people outside its borders because these interactions help U.S. officials to formulate more effective policies by helping them to know more about the people and governments that those policies seek to influence: "We will need to improve our ability to understand the concerns, perceptions, and attitudes of foreign governments and populations, as well as the ways in which our words and actions may affect allies and partners. Thoughtful engagement, communication, and collaboration with allies and partners who share our interest in fostering peace and security remain essential" (DoD *QDR* 2010: 57). Comments made by Major Lee Tan from Singapore exemplify the informational impact of military exchanges. Major Tan, a 1999 graduate of the U.S. Army Command and General Staff College, remarked upon how his exchange experience influenced his own thinking about the United States: "I've certainly developed a more informed opinion of the U.S. government and how its policies are formulated, and a better understanding of American culture in general by being immersed in it, than by watching CSPAN or local television back home" (as quoted in Dunphy 2000: 33).

A second important impact of the friendships and professional relationships established during the military exchanges is that the military epistemic community can help to extend U.S. influence across national boundaries. In addition to being a source of information, educational exchanges can be an important source of soft power. As the host organization, the U.S. government, through the U.S. military, seeks to set a good example or put forth a model that inspires and motivates the exchange participants who want to emulate this example in order to achieve a positive outcome in their home countries. Social interaction, social learning, argumentative persuasion, and normative persuasion are all factors at work here. The socialization process and resultant soft power is illustrated by remarks made by Lieutenant Colonel Andrei Demurenko, who was the first international student from Russia after the fall of the Soviet Union at the U.S. Army Command and General Staff College. LTC Demurenko was accompanied on his stay in the United States by his family, and while in the United States he wanted his children to learn skills to live successfully under a capitalist system: "I wanted to show them that here is the way things can be. Here is cleanliness, riches and beauty. But on the other hand, I want to show them that when they come to the country, nothing is thrown at you. You must work for it. . . . It is my hope that my son could work while he is here so that he would understand what work is worth, that he could learn there is value in being there on time—just the fact that he participated in the process and understands that when you work, you get paid" (as quoted in Waheed 1993: 15).

Finally, a third important impact of the military exchanges is that they are designed to teach about and foster democratic institutions and practices. Beyond extending U.S. soft power, one of the explicit goals of U.S. military educational exchanges is to foster democratic norms and support democratic governance worldwide. In their 2011 joint report to the U.S. Congress on foreign military training, the Department of Defense and the Department of State described this goal and how the military exchanges play a key role:

> Training provided under the IMET program is professional and nonpolitical, exposing foreign students to U.S. professional military organizations and procedures and the manner in which military organizations function under civilian control. The IMET program's mandatory English-language proficiency requirement establishes an essential baseline of communication skills necessary for students to attend courses. This requirement facilitates the development of valuable professional and personal relationships that have provided the U.S. access to and influence in a critical sector of society that often plays a pivotal role in supporting or transitioning to democratic government. The IMET program introduces military and civilian participants to elements of U.S. democracy such as the U.S. judicial system, legislative oversight, free speech, equality issues, and the U.S. commitment to human rights. (DoD/DoS *Foreign Military Training, Fiscal Years 2010 and 2011* 2011: II-2)

The Field Studies Program is a key component of U.S. military training programs for international students, including those international students attending professional military education courses in the United States. The program is mandated by the U.S. Department of Defense and requires each military educational institute to develop instructional courses and educational activities for their international students that provide the students with opportunities to increase their knowledge about the United States as well as democratic institutions and human rights issues.[6] The Field Studies Program at the U.S. Army Command and General Staff College is a good illustration of how this goal is pursued by one of the U.S. military's elite professional military schools. In addition to attending the regular course of study at the staff college located at Fort Leavenworth in Kansas, the international officers must participate in the Field Studies Program, which is designed to give the military exchange students an in-depth view of the life of ordinary Americans in the heartland of the United States. The yearlong Field Studies Program includes instruction and discussions on U.S. government institutions, U.S. political processes, the U.S. judicial process, the free market system, human rights, law of war, the media's role in the United States, and diversity in U.S. society (CGSC 2009). Highlights of the program for the international military officers are field trips to the U.S. capitol in Washington, DC, the Kansas state capitol in Topeka, academic institutions, local businesses, correctional facilities (that is, prisons), and mass media organizations. At the state capitol, the governor of Kansas and other officials host the officers and discuss subjects such as how citizens participate in local government and the functions of the National Guard. In a trip to the University of Kansas School of Journalism and Mass Communication the officers participate in discussions on the U.S. Constitution's First Amendment and the media's role in democracy (Kennedy 2003: 14; Lugo 2003: 12). Jim Fain, chief of the International Officer program at the Army

Command and General Staff College, has found the Field Studies Program to be instrumental in providing "insight into the American way of life, and how the military operates under civilian control" (Kennedy 2003: 14). This experience has also helped to reduce stereotypes of U.S. people and provides a more nuanced understanding of the United States and U.S. citizens than might be learned from tourist trips or through Hollywood movies and other mass media.

The three impacts discussed above are interrelated. More sources of information can help people understand each other's lives and motivations and build goodwill between them. Goodwill and increased understanding can help extend political influence, and this influence can be directed toward achieving a specific U.S. foreign policy goal such as developing democratic norms and practices in nondemocratic countries. The above joint policy statement from the Department of Defense and the Department of State illustrates that the United States uses military educational exchanges to pursue all three goals, with the explicit foreign policy objective being to build understanding, appreciation, and desire for democratic norms and practices in nondemocratic countries.

3.5. SUMMARY

The military is a key instrument of U.S. national power. It fights wars, it deters conflict, and it coerces adversaries. But, in addition to its traditional missions, the U.S. military builds its influence and soft power through a variety of interactions with other militaries around the world. Military educational exchanges are one aspect of this effort and are an important component of U.S. foreign policy that helps to build goodwill toward the United States amongst military officers around the world and, thus, extend U.S. influence globally. This chapter has described how military educational exchanges are an important "nontraditional" source of power and influence for the United States that helps the United States achieve its foreign policy goals and build its soft power.

The chapter also discussed key mechanisms that make the exchanges hosted by the U.S. military's war and staff colleges particularly conducive to successful socialization of its participants. First, the programs seek to educate elite leaders. Because military officers often influence politics within their countries, the ideas exchanged within military educational programs are important in shaping the motivations, values, and actions of political elites and the policies within the countries that they rule. In the case of democratic reforms, military officers are not just ordinary citizens. They often control the means that repressive regimes use to stay in power and can play a key role in facilitating democratic transitions. Second, the programs incorporate both argumentative and normative persuasion. The mandatory Field Studies Program employs instruction, argumentative persuasion, and normative persuasion. These mechanisms also operate within the larger course of study in which the international officers participate with their U.S. counterparts. Not only is the classroom a source of information, but it is also a living example of how U.S. military officers think and act. Third, the programs foster social interaction and social learning. The professional networks and personal friendships formed between the officers can provide cohort support and function as a military epistemic community. Both classroom learning and social learning introduce new ideas, norms, and perspectives. The next part of the book shows how these theoretical propositions play out in real life. Survey data is used to learn more about the impact of the educational exchange experience on its international participants and their countries. It shows the types of ideas, attitudes, and motivations that the international officers learn and take back home with them.

NOTES

1. The traditional and nontraditional views are Weberian ideal-types in that they are epistemological devices that describe basic characteristics of two different conceptualizations of military power.

2. For interesting analyses that challenged mainstream views see Holsti (1967), George (1969), Bull (1977), Jervis (1976), and Finnemore (1996).

3. Georgi Arbatov was (and still is) highly esteemed as an expert analyst of U.S. politics and military policy. He is a member of the Russian Academy of Science and during the Cold War served as the director of the Institute for the Study of the USA and Canada, the primary organization that provided advice to the Soviet leadership on the foreign policy of the United States. Arbatov served as an advisor to Soviet leaders including Stalin, Khrushchev, Brezhnev, Andropov, and Gorbachev.

4. The six countries of the Gulf Cooperation Council were initially invited to join. Bahrain, Kuwait, Qatar, and the United Arab Emirates have joined. Saudi Arabia and Oman so far have not joined. The program is now open to all interested Middle Eastern countries as well as the Palestinian Authority.

5. Members of the Desert Storm coalition were Australia, Bahrain, Bangladesh, Canada, Czechoslovakia, Denmark, Egypt, France, Germany, Greece, Honduras, Hungary, Italy, Kuwait, Morocco, Netherlands, New Zealand, Niger, Norway, Oman, Pakistan, Poland, Portugal, Qatar, Saudi Arabia, Republic of Korea, Senegal, Spain, Syria, Turkey, United Arab Emirates, United Kingdom, and the United States.

6. Each Field Studies Program must provide participants with "awareness and understanding of the following facets of the American democratic way of life"—human rights, law of war, international peace and security, U.S. government institutions, U.S. political processes, U.S. judicial system, U.S. economy and the free market system, U.S. media and the role of the free press, U.S. health and human services, and diversity within the United States (DoD 2006).

II

IMPACT OF MILITARY EDUCATIONAL EXCHANGES ON PARTICIPANTS

4

The History of Educational Exchanges at U.S. War and Staff Colleges

Military educational exchanges encompass a variety of activities through which participants learn not only about military subjects but also about their colleagues' own experiences. Within these programs, military officers are socialized to the ideas, doctrines, and operating procedures taught at the schools. Part of the learning experience is attending lectures, part is seminar discussions with fellow students, and part is informal interactions both on- and off-duty that the officers have with each other. For international participants, the exchanges also offer the opportunity to live in another culture.

This chapter describes the professional military exchanges hosted by the U.S. military's war and staff colleges. The chapter begins by placing these exchange programs within the broader context of U.S. military doctrine and U.S. military strategies and shows the importance given to these programs by U.S. government and military officials. The U.S. government's primary program that provides grant funding for a variety of education and training programs for foreign military personnel is called the International Military Education and Training (IMET) program. The chapter discusses the goals, nature, scope, and costs of IMET programs and explains the extent to which the international officers who participate in the educational exchanges at U.S. military war and staff colleges are encompassed under IMET. Finally, the chapter provides detailed information on the military exchange programs at each of the U.S. military's war and staff colleges. While the courses of study at these schools share many similarities, there are also some rather substantial differences between them. Each school's unique history and mission is discussed with a focus on how, when, and the extent to which foreign military personnel have been incorporated into their student bodies. Each school has a substantial number of international alumni who have gone on to hold some of the highest positions in their country's armed forces or government, and the chapter provides data on how many foreign graduates from each school have achieved these very influential positions.

4.1. THE ROLE AND VALUE OF MILITARY EDUCATIONAL EXCHANGES

For over one hundred years military personnel from around the world have been invited to participate in numerous technical training and professional education programs hosted by the U.S. military. In fiscal year 2010 alone approximately 67,100 students from 159 countries participated in U.S.-sponsored military training under numerous security cooperation initiatives (DoD/DoS *Foreign Military Training* 2011: ii). As discussed in the previous chapter, recent U.S. military operations in Iraq and Afghanistan, as well as membership in longstanding alliances such as NATO, have demonstrated the importance of developing and sustaining military-to-military relationships between the U.S. military and its allies and potential partner nations. Not only can these professional military networks increase interoperability in times of war, but even more important, they build confidence, trust, and mutual understanding that can prevent disagreements from escalating to armed conflict and war in the first place. Perceptions of the value of military-to-military exchanges have changed in recent years as their value in building cross-border linkages has been increasingly recognized by U.S. officials. As one example of the impact of changing perceptions, in a rather astounding doctrinal change, the U.S. Air Force now considers "building partnerships" a core competency equivalent to its historical raison d'être of air superiority.

Military commanders have highlighted the importance of building partnerships and engaging future leaders within their regional areas of operation. The former commander of U.S. Pacific Command, the United States' largest regional command, described security assistance activities that focused on building partner capacity as "among our most powerful" strategies because "these programs *expose future leaders of other countries to our values and culture through education and training*; present opportunities for nations to purchase U.S. military equipment that enables greater interoperability in our combined operations; and *provide engagement opportunities for our best ambassadors, our young servicemen and women, to develop relationships with the region's military personnel and general populations*" (Willard 2010, italics added). Likewise, the former commander of U.S. Central Command who oversaw, arguably, the most volatile region in the world testified before the U.S. Senate Armed Services Committee that as a nation the United States must "look beyond just the traditional application of military power and integrate all elements of national power" (Mattis 2011: 44). In this effort, military-to-military engagement activities serve as both a cost-effective investment and "a critical component" of U.S. strategy (Mattis 2011: 33). General Mattis made the key point that while military-to-military engagement might not have immediately observable and measurable effects, these activities that have been "decades in the making" have helped partner nations to become "the professional forces they are today—and in the process made our forces better as well" (Mattis 2011: 4). One interesting illustration of the value that Central Command places on military-to-military exchanges is the composition of its own headquarters in Tampa, Florida, which hosts 193 foreign personnel from fifty-eight different allied countries. Mattis (2011: 34–35) noted that these exchange participants "make significant contributions to [U.S.] efforts, and receive invaluable experience interacting with both US forces and our allies."

Within the broader defense establishment, former U.S. Secretary of Defense Robert Gates in the 2010 *Quadrennial Defense Review Report* identified security assistance as one of the key capabilities that the United States needed to strengthen. He particularly noted that exchanges between U.S. soldiers and their international colleagues are key to achieving U.S. national security goals in promoting international peace and cooperation:

> Since the United States assumed the role of a leading security provider after the end of World War II, DoD has worked actively to build the defense capacity of allied and partner states. Doing so has also given the U.S. Armed Forces opportunities to train with and learn from their counterparts. These efforts further the U.S. objective of securing a peaceful and cooperative international order. Security cooperation activities include bilateral and multilateral training and exercises, foreign military sales (FMS) and financing (FMF), *officer exchange programs, educational opportunities at professional military schools, technical exchanges*, and efforts to assist foreign security forces in building competency and capacity. In today's complex and interdependent security environment, these dimensions of the U.S. defense strategy have never been more important. (DOD *QDR* 2010: 26, italics added)

The above official statements all underscore the importance of building and supporting military partnerships between the United States and other nations within the highest echelons of the U.S. military and the U.S. Department of Defense. There are many programs that contribute to this goal and some of the most important are the professional military educational exchanges hosted by the United States through International Military Education and Training (IMET).

4.2. INTERNATIONAL MILITARY EDUCATION AND TRAINING (IMET)

International Military Education and Training (IMET) is the U.S. government's centerpiece program that offers funding for military training and education for military personnel from around the globe. The U.S. Congress established IMET through the International Security Assistance and Arms Export Control Act of 1976. The U.S. Department of State and the U.S. Department of Defense share responsibility for the program; IMET courses are funded through the U.S. Department of State but the courses are run by the U.S. Department of Defense. IMET is a grant program. The Foreign Assistance Act of 1961, as amended (Public Law 87–195), Section 544, authorizes the president to "provide for attendance of foreign military personnel at professional military education institutions in the United States (other than service academies) without charge" although the president may require "an equitable contribution of support and services from each participating country" (U.S. Congress 2010: 294–95). The legislation prohibits the providing of grant assistance funding to "certain high income foreign countries" (U.S. Congress 2010: 294–95). This is important to note because it means that not all of the international officers who attend education and training courses in the United States can be tracked solely through IMET databases. This does not mean that wealthier countries (such as Austria or Saudi Arabia) do not participate; it merely means that their personnel are funded through different means. A wealthier country might also purchase participation for its military officers, for example, through U.S. Foreign Military Sales activities.

Although not stated in such terms, the central goal of IMET is to socialize foreign military personnel through interactions with U.S. military personnel and within education programs on U.S. military installations. The socialization has many aspects with three primary goals: to professionalize foreign military forces, to improve interoperability with U.S. military forces, and to increase respect for democratic values and human rights. In the words used by U.S. officials, the overall objectives of the IMET programs are to

1. further the goal of regional stability through effective, mutually beneficial military-to-military relations that culminate in increased understanding and defense cooperation between the United States and foreign countries;
2. provide training that augments the capabilities of participant nations' military forces to support combined operations and interoperability with U.S. forces; and
3. increase the ability of foreign military and civilian personnel to instill and maintain democratic values and protect internationally recognized human rights in their own government and military. (DoD/DoS *Foreign Military Training* 2011: II-2)

Recent U.S. legislation that increased U.S. government reporting on security assistance programs, such as IMET, is another indication of the increasingly valuable political role played by this "nontraditional" aspect of U.S. military power. Through legislation, the U.S. Congress has sought greater efforts by U.S. officials to track the foreign personnel who attended U.S. training and education courses and, concomitantly, more information on the subsequent careers of the foreign personnel who have attended IMET programs. The Foreign Relations Act of 2003 required the U.S. Department of Defense to develop and maintain a database with a record for every foreign military person or defense ministry civilian who had participated in IMET education and training activities beginning on 1 January 2001. The U.S. Secretary of State is now required to submit a report to the U.S. Senate and U.S. House of Representatives on any IMET graduate who is found to have violated internationally recognized human rights (U.S. Congress 2010: 296–97). As noted in chapter 1, this database is still a work in progress with continued "weaknesses in monitoring IMET graduates and assessing IMET program effectiveness" (U.S. GAO 2011: 26). Nevertheless, the effort to gather and maintain this detailed database illustrates the importance to the U.S. government of developing cross-border military networks and maintaining an international community of military partners.

IMET programs are extensive and varied, but they are also relatively inexpensive. They are routinely hailed as "a low-cost, highly effective component of U.S. security assistance" (DoD/DoS *Foreign Military Training* 2011: II-2). In 2010 under IMET, 7,000 foreign military and civilian personnel from 136 countries[1] received training in more than 4,000 formal courses at approximately 150 U.S. military schools and installations for a total cost of around $96.7 million[2] (DoD/DoS *Foreign Military Training* 2011: II-2) or about 0.2% of the budget of the State Department.[3] Operating budgets for the local units that host the international military officers on U.S. military bases are also small. The International Officer School at Maxwell Air Force Base in Alabama is a good example. The International Officer School is the central hub for all international exchange officers attending the schools of Air University. It provides substantial support to the international exchange students throughout their stays—from providing the initial welcome and orientation for the officers, teaching the initial preparatory classes, organizing professional meetings, field trips, and social events, and overseeing support and administrative aspects of the officers' programs—all on an annual budget of $100,000 (Brown 2011). In fact, as will be discussed later, the program could not be sustained in its current form without the help of many volunteers from the local civilian and military communities.

Most programs funded through IMET are short in duration and aimed at technical training; however, IMET also sponsors many of the international officers who attend the yearlong professional military education programs at the U.S. military's war and staff colleges. In terms of IMET program costs, in FY2010 professional military education (PME) accounted for nearly 50% of IMET funding with English language training comprising 13% and technical training accounting for 11% of the budget. As these budgeting priorities indicate, the professional military education exchanges hosted by the war and staff colleges are very important in terms of accomplishing the overall objectives of IMET. One of the reasons for this is that these exchanges build networks of influence. In terms of socializing effects, they have the highest probability of long-term impact on both the individual participants and their countries. There are a number of reasons why this is. First, the war and staff colleges host senior military officers who are specially selected and likely to achieve high rank in their home countries. As officials at the U.S. State Department emphasized, "Highly sought-after PME courses are key in ensuring that we can attract qualified individuals destined for future leadership positions and [they] are a critical piece of imparting our values" (U.S. GAO, Appendix III Comments from the Department of State 2011: 39). All of the schools have numerous international (and U.S.) graduates who have achieved the highest ranks in their military services; a few have even become heads of state. Second, the programs are 10–12 months in duration providing intensive and extensive exposure to U.S. society as well as the best opportunities for forming enduring friendships and professional relationships. Third, most of the officers bring their spouses[4] and children with them. The family's experience is often key to the officer's positive experience.

The professional military education exchange programs share similar goals across the various war and staff colleges. These goals are mandated by the U.S. Congress and shaped by each of the military services to support their own warfighting needs. Nevertheless, the international exchange programs at the war and staff colleges vary in their historical development, student body size, and proportion of international participants. The officials who oversee the programs at each school are also influential in shaping how the programs are run and their level of success. For example U.S. program administrators (and their spouses) often manage and help organize volunteers to support unfunded, but very important, aspects of the international officers' stay in the United States.

Before discussing the unique history of each of the U.S. war and staff colleges, the next section looks at general characteristics that span the programs including the selection process for attendance, the importance of these schools to the careers of U.S. military officers, and how the international officers and their families are incorporated into the programs.

4.3. U.S. WAR AND STAFF COLLEGES AND THEIR STUDENTS

Within the U.S. military, each of the four military services[5] has its own war college and its own staff college devoted to educating its rising senior leaders.[6] At the national level the Department of Defense also runs several schools in Washington, DC, that are focused on national-level decision-making, the integration of the military capabilities of all of the services, and the fostering of inter-agency cooperation between the military and relevant agencies of the U.S. government. For U.S. officers the selection process for in-residence attendance at these elite military schools is very competitive. The officers are chosen by special selection boards based on their overall military record.[7] For international officers, the process is more complicated. Each U.S. military service holds a DoD-level selection board that decides what countries will be offered slots at their schools. Generally, only one slot is offered to any one country at any one school. The goal is to have a balanced worldwide representation of countries throughout the military school system. The schools then send out invitations to the selected countries, and the countries are invited to nominate candidates that meet the school's criteria for such things as rank, length of military service, military experience, educational degree, and English fluency. Nominees are vetted by the local U.S. embassies, and the final choice is made by the officer's home country.

For the international officers, their selection to attend a U.S. war or staff college is clear affirmation of their importance within their own militaries and their own potential to rise to high-ranking positions back in their home countries. For example an international officer from the Americas noted that graduation from a U.S. military school would have a "strong" impact on his military career; he went on to elaborate, "It is a unique opportunity that implies a high level of trust from my superiors."[8] An officer from Asia echoed this sentiment; he also thought that his graduation from one of the staff colleges would have a "big impact" on his military career. He then explained that when he returned to his home country after graduation his "military and civilian leaders will have high expectations for me to excel in every task or assignment." Today, countries send their brightest and best officers to attend U.S. war and staff colleges. This was not always the case. In the early 1970s around 50% of the international officers were selected by their countries based on patronage or family ties; however this figure has dropped to less than 25% in the past decade and continues to decrease (Homan 2011). Countries spend money and use resources to send their officers to the United States, and as resources have gotten more and more constrained, the leaders in these countries recognize that they get more value for their money by sending the best and brightest officers rather than those with family or political ties (Homan 2011).

The majority of officers that attend staff colleges are in the rank of major (or the equivalent naval rank of lieutenant commander). They are at the midpoint of their military careers. They are generally in their mid-thirties and have served in the military as an officer for roughly twelve years. For U.S. officers, graduation from a staff college is a discriminator for further promotion; other things being equal, those military officers who graduate from a staff college are more likely to be promoted to colonel and are more likely to be selected to attend a war college. The officers attending war colleges are lieutenant colonels or colonels (and the equivalent naval ranks of commander and captain). They are generally in their mid-forties and have roughly twenty years of military service as an officer. In many ways, U.S. war college graduates are seen as the elite of the U.S. officer corps, and they are considered very competitive for promotion into the general officer ranks and for selection to important operational command positions. The same is also true for the international officers as some of the statistics previously discussed show.

Each of the war and staff colleges welcomes a new class of students in the summer of each year.[9] The U.S. students usually arrive in late July or early August. The international officers typically arrive earlier and attend orientation programs. For example, at Air University, the international students arrive two months earlier in order to attend a seven-week orientation program at the International Officer School that includes introductory classes as well as training mandated through the Field Studies Program.[10] During the summer while the officers are attending their formal training, their families may also participate in their own orientation programs. Under Title 10 of U.S. federal law, U.S. government funds cannot be used to run programs for families, so these programs are organized and run by volunteers. For example, the International Family Orientation Program at Air University is a two-week program that provides information on the local community for spouses and recreational trips in the local area for spouses and their children. At Air University, volunteers also manage the International Family Loan Closet Program where international families can check out household items such as dishes and kitchen utensils and small appliances. Volunteers also teach a six-week long voluntary English language course for all of the international officers' spouses. Although volunteers run all these programs, the programs are beneficial for the international officers because they help the officers focus on their own studies (Homan 2011).

All of the war and staff colleges run sponsorship programs for their international participants. Upon arrival at the war or staff college, each international officer is assigned at least one sponsor within the local community. The sponsor program focuses on helping the international officers adapt to life in the United States through informal socialization to political, social, economic, and other cultural aspects of life in America. The sponsors help the officers with the many tasks of organizing their lives in the United States, from setting up utilities to enrolling children in schools. The sponsors invite the officers and their families to dinners and social events in their homes and provide orientations to the community and community events. The U.S. Army Command and General Staff College has a particularly robust sponsor program where international students are assigned a military sponsor and two civilian sponsors. The military sponsor is a U.S. soldier assigned to Fort Leavenworth. For the civilian sponsors, one is from the local community of Leavenworth and the other is from the larger Kansas City metropolitan area. The military sponsor program was established in 1948 and is under the direction of the Command and General Staff College; however, the civilian programs are not administered by the military. For the civilian programs, the local Leavenworth program was established in 1952 and is run by the Leavenworth-Lansing Area Chamber of Commerce (LV MAG 2009: 25). The Kansas City program is run by the Greater Kansas City Chapter of People to People International (GKC People to People 2010).

The U.S. military encourages the international officers to bring their families with them to the United States. Many of the officers do choose to do so and are accompanied by their spouses and children. Housing varies with location. Most international officers who are accompanied by their families do not live on the military base, but live within the local community in homes or apartments either rented by them or by their home country government. Unaccompanied officers may choose to live in an apartment on the military base or within the local community. Many of the unaccompanied officers choose to live on the military base for convenience, to save money, and to be near to each other.

Each war and staff college has its own unique history that includes how, when, and the extent to which it has incorporated international officers into its programs. In the next section, the history of international officer attendance at each of the U.S. military's war and staff colleges is discussed to give the reader an idea of the scope and magnitude of each school's program. The following sections describe how international officers have been incorporated into each school's curriculum, the composition of recent graduating classes, and the success of international graduates in achieving appointment to higher ranks. The first section looks at the U.S. Army Command and General Staff College because its history of integrating foreign officers into its courses is the longest of any of the schools. The U.S. Army is also unique in that its staff college is geographically separated from its war college. The other military services have all chosen to colocate their war and staff colleges.

4.3.1. Army Command and General Staff College, Fort Leavenworth, Kansas

The long tradition of hosting international officers at U.S. elite military schools began at the U.S. Army Command and General Staff College when, in 1894, Swiss officer Henri LeComte became the first international student to attend courses there (then called the U.S. Infantry and Cavalry School).[11] According to historian John Reichley (1994), Henri LeComte's father, also a Swiss military officer, had served in the United States as an observer of U.S. Army operations and aide-de-camp for Major General McClellan during the U.S. Civil War. The elder LeComte had used his influence to gain a place in the U.S. Military Academy at West Point's Class of 1893 for his son, Henri. After graduation from West Point, Henri LeComte went on to study with the U.S. Army at Fort Leavenworth for four months. LeComte's U.S. military schooling and connections with U.S. military officers would later prove useful. During World War I, he worked with former West Point and Leavenworth classmates in the headquarters of the American Expeditionary Force headquarters in March–April of 1918. Although LeComte attended classes at Leavenworth, he did not complete the entire two-year course of study.[12] The first recorded international graduates of the U.S. Army Command and General Staff College were Lieutenants J. Landa and L. Monter from the Mexican Army who graduated in 1908, followed by a Puerto Rican officer in 1910, and two Cuban officers in 1912 (Reichley 1994).

Since these early years, the attendance of foreign officers at the U.S. Army's staff college has been greatly expanded. As of 2012, more than 7,300 international students had graduated from the Command and General Staff College. Of these, more than half have obtained the rank of general, and 241 officers from sixty-seven different countries have become heads of their militaries or heads of state (CGSC *Hall of Fame* 2011).

The curriculum at the U.S. Army Command and General Staff College focuses on developing leaders not only with expertise in land warfare but also the capability of operating with the other U.S. military services, with relevant civilian agencies, and with foreign military personnel. In the classes graduating in 2012, 97% of the U.S. Army officers had combat experience; 72% had two or more combat tours (Sarles 2012). Distinguished U.S. graduates include former U.S. Secretary of State Colin Powell and one of the most recognizable commanders from the recent wars in Iraq and Afghanistan, General (now retired) David Petraeus. Not only does the Command and General Staff College have the longest running exchange program, it also has the largest student body of all the war and staff colleges. At any one time there are approximately 1,400 students at the Command and General Staff College.

Prior to the start of classes, international officers complete a two-month orientation course at Leavenworth. They are then fully integrated into seminar groups with

Table 4.1. International Students U.S. Army Command and General Staff College 2010–2012 Classes

	2010–01	*2010–02*	*2011–01*	*2011–02*	*2012–01*	*2012–02*
Total students	1046	399	1040	336	1055	336
International officers	67 (6%)	46 (12%)	69 (7%)	47 (14%)	69 (7%)	46 (14%)

SOURCES: Fort Leavenworth Public Affairs Office (2009, 2010, and 2011) and Sarles (2012)

their U.S. counterparts. The college runs two classes per year; the main class enters in August and graduates in May (designated "-01") and the other class enters in February and graduates in December (designated "-02"). The course of study is identical for both classes. The most recent summer graduating classes have had over 1,000 students with approximately 7% of graduates being international officers. The winter classes have a smaller student population overall, but international students make up roughly 14% of these students. The figures for the most recent graduating classes are shown in table 4.1.

4.3.2. Army War College, Carlisle Barracks, Pennsylvania

International officers have also been integrated with their U.S. counterparts in the student body of the U.S. Army War College, which hosts more senior level officers. The Army War College was established in 1901 at Fort McNair in Washington, DC, and its first class of U.S. officers graduated in 1904. The Army War College was originally established to study lessons learned during the Spanish-American War and to provide advice to the Army General Staff and the U.S. president.[13] Today its mission is to prepare its graduates to assume the highest levels of command; and many have achieved just that, with President Dwight Eisenhower as one of the Army War College's most distinguished U.S. graduates. The War College is considered the Army's "most senior and most prestigious academic institution" (Williams 2010).

Unlike the Army's staff college that has had a very stable existence at Fort Leavenworth, the Army War College has undergone several significant disruptions. It was closed during the First World War, reopened in 1919, and was again closed during the Second World War (U.S. Army War College 2010). After much debate on its relevance in the post-war years, the Army War College was again reopened in 1950 and temporarily moved to Fort Leavenworth in Kansas. In 1951 it was relocated to its present location at Carlisle Barracks in Pennsylvania where it has remained ever since (U.S. Army War College 2010; NDU *History* 2010).

Officials at the Army War College initially resisted integrating international officers into classes with their U.S. counterparts. After World War II there were several proposals put forth to incorporate international officers into the Army War College's program, but they were rejected for a variety of reasons, including concerns with the handling of classified information, the inhibiting effect that foreigners might have on guest speakers, and the spending of U.S. taxpayer dollars on foreign rather than U.S. officers (Ball 1983: 406, 448, 475). In his superb history of the U.S. Army War College, historian Harry Ball (1983: 475) noted, "Of all the schemes contemplated to bring different points of view to Carlisle Barracks, none was more difficult to implement than one involving having officers from armies of other nations at the War College on a resident basis." According to Ball (1983: 455–83), the successful incorporation of foreign officers into the Army War College program was primarily due to the efforts of Major General DeWitt C. Smith, Jr., who served as Commandant of the Army War College from 1974 to 1977. Smith was convinced (much like Clausewitz had argued) that foreign officers would bring new views into the War College that would ultimately benefit the U.S. officers. He and a few of the faculty members believed that the professional and social interactions amongst all participants would also foster greater international understanding.

The Army War College's International Fellows Program was eventually approved in 1976. The first six international fellows graduated in the summer of 1978; they were from Australia, Canada, West Germany, Japan, South Korea, and Mexico. Within two years there were enough international fellows to assign one to every seminar group with the graduating class of 1980 including thirty-two international fellows representing twenty countries (Ball 1983: 476). Since these early beginnings, the International Fellows Program has continued to expand. The most recent graduating classes have continued to grow in size, and the proportion of international fellows in each class has also increased as can be seen in table 4.2. The class of 2012 is the largest ever, and its international fellows hail from around the world.[14] As of January 2011, thirty-two international fellows from the U.S. Army War College had achieved the highest positions in their armed forces or an equivalent high-ranking position in a multinational organization (Gregg 2011).

One sure sign of the success of the International Fellows Program is that the Army War College plans to increase the number of international fellows dramatically in the next few years to eighty or about 20–25% of the student body (Associated Press 2010). Ball's assessment of the success of the International Fellows Program seems to ring just as true today as it did in 1983: "The benefits to the War College were significant; a dimension had been added to the educational environment that broadened substantially the academic outlook of the students and the faculty. Some of the severest critics of the idea became the strongest supporters of the International Fellows Program" (Ball 1983: 476).

Table 4.2. International Students
U.S. Army War College
2010–2012 Classes

	2010	*2011*	*2012*
Total students	322	336	368
International officers	50 (16%)	49 (15%)	67 (18%)

SOURCES: For the class of 2010 (Baltos 2010), for the class of 2011 (Zimmerman 2011), and for the class of 2012 (Cress 2011)

4.3.3. Air University, Maxwell Air Force Base, Alabama

Air University at Maxwell Air Force Base in Montgomery, Alabama, is the U.S. Air Force's central hub for professional military education. Air University was literally built upon U.S. Army institutions that were developed in the aftermath of World War I. During this era, there was no separate military service for air forces. Both the U.S. Navy and the U.S. Army carried out research and worked on developing effective strategies for using airplanes in military operations. However, it was the U.S. Army from which the doctrines and institutions of the U.S. Air Force were inherited. Air University had its genesis in the decision of the U.S. Army to relocate its Air Corps Tactical School (ACTS) to Maxwell Field (as it was then called) in 1931. ACTS was the first U.S. military school devoted to the professional military education of airmen and developed much of the air doctrine used by U.S. Army Air Corps fliers in World War II. In 1946 after the war was over, the Army established Air University to continue work on the development of airpower doctrine. In 1947, the U.S. Air Force achieved status as its own military service, and it retained much of what had belonged to the Army Air Corps, including Air University. Today, Air University encompasses several research institutes as well as numerous schools for both officers and enlisted personnel at various stages in their careers.

The Air War College and the Air Command and Staff College are the U.S. Air Force's premier institutes for the education of its future leaders. International officers have been integrated into both programs since the 1950s. During its earlier years, international officers only attended the first half of the programs and graduated in February because the second half of the programs dealt with classified information that only U.S. students could have access to (McCarthy 2011). This changed in 1990–1991 when the U.S. Air Force began to allow the international officers to participate in the entire yearlong course of study. Today, very little of the material discussed in either school is classified (McCarthy 2011). Both schools, similar to the other war and staff colleges, welcome a new class of students in the summer of each year.

The International Officer School at Maxwell Air Force Base is the central organization that oversees all aspects of the international officers' stays at Air University. Much like the programs at the other war and staff colleges, the international officers typically arrive earlier than their U.S. counterparts in order to attend an orientation program. At Air University the newly arrived officers attend a seven-week-long Field Studies Program that helps to prepare them for life in the local community and for their educational courses. The program includes exposure to many facets of U.S. society as well as trips around the base, in the local area (Montgomery and Atlanta), and to Washington, DC (Homan 2011). While the officers are attending their mandatory orientation program, the officers' spouses and children may also participate in the International Family Orientation Program that is run by volunteers, many of whom are spouses of U.S. officers and other personnel who work on the base. This two-week program helps the entire family settle into their new life by providing information on the local community, recreational trips for the children, and informa-

Table 4.3. International Students
Air War College and Air Command and Staff College
2010–2012 Classes

	Air War College		
	2010	*2011*	*2012*
Total students	238	240	242
International officers	45 (19%)	44 (18%)	46 (19%)
	Air Command and Staff College		
	2010	*2011*	*2012*
Total students	506	509	514
International officers	76 (15%)	79 (16%)	75 (15%)

SOURCES: For AWC: for 2010 (Maxwell AFB 2009), for 2011 (Deichert 2011), for 2012 (AWC Demographics 2012); for ACSC: for 2010 (Wright 2010), for 2011 (Kratzer 2011), for 2012 (ACSC GOE 2012)

tion on daily living (such as how to open a bank account or call 911), to help reduce culture shock for the officers' families (Homan 2011).

At Air University, the international officers are integrated with their counterparts in the Air Force's war and staff colleges and form a substantial cohort each year. The statistics for attendance in the most recent classes at each of the schools are shown in table 4.3.

Like their international counterparts in the U.S. Army schools, the international officers at Air University are accomplished leaders who go on to hold some of the highest ranks in their home countries. In fact, the international graduates are consistently promoted at higher rates than their U.S. counterparts: 2.5% of the international officers who graduate from the Air War College or the Air Command and Staff College go on to achieve the highest rank in their country's military. In comparison, only 1% of U.S. officer graduates will achieve a similar rank in the U.S. military (Homan 2011). As of February 2012, there have been 413 international graduates of Air University who have been promoted to chief of their country's air forces or an equivalent high ranking civilian or military position. These esteemed graduates represent eighty-eight different countries (AU IOS 2012).

4.3.4. Marine Corps University, Quantico, Virginia

The U.S. Marine Corps University in Quantico, Virginia, is home to the Marine Corps War College, the Marine Corps Command and Staff College, and the School of Advanced Warfighting. The U.S. Marines have historically had numerous formal education courses for their officers and enlisted personnel. These date back to 1891 when the U.S. Marine Corps established its first in-residence course, called the School of Application at Marines Barracks in Washington, D.C. (USMC 2010: 1). The School of Application was renamed the Officers Training School in

1909 and moved to Quantico in 1917. At this time the course of study focused on training officers who were deploying to fight in World War I. Deficiencies that were identified during operations in World War I spurred the further development of a system of formal education for marines on tactics and weapons use that resulted in the establishment of the Field Officers Course and the Company Grade Officers Course in the early 1920s. It was also during the 1920s that the U.S. Marine Corps embraced the amphibious warfare mission and Marine Corps training began to focus on amphibious operations. These three early schools, Officers Training School, Field Officers Course, and Company Grade Officers Course, were the foundation of the U.S. Marine Corps professional military education system and were eventually transformed into the current-day Marine Corps University (USMC 2010: 1–3).

Marine Corps University was not officially founded until 1989, when the Marine Corps consolidated fourteen schools at Quantico under one organization (MCU Foundation 2010). The new organization included the Marine Corps Command and Staff College, which was formerly known as the Field Officers Course. It was not until 1991 that the Marine Corps War College (first called the Art of War Studies program in 1990) was formally incorporated into Marine Corps University (Blair 2002; USMC 2010). The statistics for international officer attendance at the U.S. Marine Corps' war and staff colleges in recent years are shown in table 4.4.

Table 4.4. International Students
Marine Corps War College and Marine Corps Command and Staff College
2010–2012 Classes

	Marine Corps War College		
	2010	*2011*	*2012*
Total students	27	27	27
International officers	3 (11%)	3 (11%)	2 (7%)
	Marine Corps Command and Staff College		
	2010	*2011*	*2012*
Total students	198	199	205
International officers	26 (13%)	26 (13%)	28 (14%)

SOURCE: MCU 2012

4.3.5. Naval War College, Newport, Rhode Island[15]

The Naval War College was established in 1884 and today consists of six separate colleges.[16] Unlike the other services' war and staff colleges, historically the U.S. Navy did not integrate international officers into their main programs but developed separate programs for them. At the Naval War College there were two resident programs that were open to international officers: the Naval Command College and the Naval Staff College. As of 2012, officers from 129 different countries had attended one of these programs (U.S. Naval WC *Programs* 2012). Like the other services, many of the graduates have gone on to achieve high rank and important positions.

At the more senior level the Naval Command College has been hosting international officers since 1956. The program of study lasts an entire year. Although the international students are in a separate program in a separate school, in 2004–2005 they were fully integrated with their U.S. counterparts in the academic classroom (U.S. Naval WC *Programs* 2012). The number of international officers attending the Naval Command College program has grown over the years. In 2012 there were a total of 45 international naval officers. In comparison, the College of Naval Warfare, which is the equivalent U.S.-only school, had 175 U.S. military officers and 22 U.S. civilians. So, international participants accounted for roughly 19% of personnel attending the senior-level naval programs (U.S. Naval WC *Fast Facts* 2012). As of July 2011, 1,905 international officers had graduated from the Naval Command College; 942 (49%) of these graduates had obtained flag rank; 209 (11%) had served as the chief of their country's naval forces; and twelve were currently serving as chief of their country's naval forces (U.S. Naval WC *Fast Facts* 2012).

At the more junior level, the number of international participants has also been growing in recent years. In 2012 there were 42 international naval officers attending the Naval Staff College. In comparison, the College of Naval Command and Staff, which is the equivalent U.S.-only school, had 269 U.S. military officers and 13 U.S. civilians attending its courses. So, international participants accounted for roughly 13% of personnel attending staff college level courses (U.S. Naval WC *Fast Facts* 2012). The curriculum for international officers at the Naval Staff College differs from the other U.S. military services' staff colleges. The school offers both a yearlong program and a six-month short course where international officers are only partially integrated with their U.S. counterparts. The Naval Staff College has been hosting international officers since 1972, but at that time international officers were only allowed to attend a six-month program and did not participate in classes with their U.S. counterparts. It was not until 2004–2005 that international officers were offered a yearlong course comparable to the U.S.-only version and were integrated with their U.S. counterparts in the comparable U.S.-only school (U.S. Naval WC *History* 2012). As of July 2011, 2,027 international officers have graduated from the Naval Staff College; 329 (16%) of these graduates had obtained flag rank; and 122 (6%) had served or were currently serving as the chief of their country's naval forces, of which nineteen were currently serving as the chief of their country's naval forces (U.S. Naval WC *History* 2012).

4.3.6. National Defense University, Fort McNair, Washington, DC

The National Defense University (NDU) at Fort McNair in Washington, DC, encompasses the most elite of the U.S. military's professional military schools.[17] Civilian attendees account for a greater proportion of the students in these national level schools than at any of the military service schools. As an example, in 2010, 55% of the graduates from NDU schools were U.S. military officers, 35% were U.S. federal government civilians, and 15% were international students (Steele 2010: 1). There are five separate colleges at NDU: College of International Security Affairs, Dwight D. Eisenhower School for National Security and Resource Strategy (formerly known as the Industrial College of the Armed Forces),[18] Information Resources Management College, Joint Forces Staff College, and National War College. These schools are under the direction of the U.S. Chairman of the Joint Chiefs of Staff.

The National Defense University's schools seek to train future national defense leaders, both military and civilian, rather than develop particular service-related military expertise. While the individual U.S. military services can claim a long history of educating their own officers in their own doctrines, this is not true at the national level where joint education that focuses on the integrated use of all of the military's capabilities as well as inter-agency cooperation is a relatively recent phenomenon. The United States first began joint education during World War II, but in the following years it was mostly neglected as the services preferred to groom their own leaders in their own service doctrines. Joint education gained stature following military reforms that began in the mid-1980s with the passage of the Goldwater-Nichols Department of Defense Reorganization Act of 1986.[19] Not only did the stature of joint education receive increased emphasis, so too did the importance of working effectively with allies and partner nations. It was during this period, in June of 1984, that the U.S. Joint Chiefs of Staff formally established the International Fellows Program at the National Defense University. During the initial year of the program there were only six international fellows, but at the end of the year support for international participation was overwhelming with 85% of students and faculty endorsing it. By 1988, international fellows were fully incorporated into the student bodies of the National War College and the Industrial College of the Armed Forces (NDU ISMO *History of IF Program* 2012). The National War College and the School of National Security and Resource Strategy are the two national-level schools focused on joint warfighting and inter-agency coordination that are historically considered the equivalents of the individual military services' war colleges.

The National Defense University's premier institute, the National War College, educates senior leaders on the warfighting aspects of joint operations. It is considered by many to be the most prestigious of all U.S. military schools. The National War College was established in the aftermath of World War II in order to educate selected officers on the capabilities of all military services and the principles of joint and combined military operations. It was literally (physically and intellectually) built on the foundations that the U.S. Army had established prior to World War II and during the war. In 1946, the National War College welcomed its first class of students into the same building on Fort McNair that had previously housed the U.S. Army's War College (NDU *History* 2010). The first class graduated in 1947 and consisted of one hundred U.S. personnel and, notably, included six foreign observers (NDU *History* 2010).

The School of National Security and Resource Strategy (formerly called the Industrial College of the Armed Forces) focuses on educating senior leaders on national resource management, materiel acquisition, and logistics. It has a longer history than its sister NDU schools. It was founded in 1924 to study mobilization, logistics, and wartime procurement problems that had hindered U.S. and allied efforts in World War I. At this time it was known as the Army Industrial College. It closed during the early years of World War II and reopened in 1943 in the Pentagon. In 1946 it was renamed the Industrial College of the Armed Forces and moved to its current location near the National War College on Fort McNair (ICAF 2010). In 2012 it was again renamed, this time, the Dwight D. Eisenhower School of National Security and Resource Strategy, but remained unchanged in other aspects.

International officers are a significant portion of the student bodies in these schools at the National Defense University, as can be seen in the statistics from the more recent classes shown in table 4.5. As of February 2012, thirty-three international graduates of the National Defense University had gone on to hold top positions in their armed forces or in multinational forces[20] (NDU ISMO 2012).

Table 4.5. International Students
National War College and School of National Security and Resource Strategy
2010–2012 Classes

	National War College		
	2010	*2011*	*2012**
Total students	221	221	221
International students	31 (14%)	34 (15%)	34 (15%)
	School of National Security and Resource Strategy		
	2010	2011	2012*
Total students	320	318	320
International students	23 (7%)	28 (9%)	26 (8%)

* Numbers based on enrollment as of 1 October 2011

SOURCE: National Defense University, 1 March 2012 correspondence

4.4. SUMMARY

International officer exchange programs at U.S. military war and staff colleges have a rich history and influential legacy of educating senior military leaders in the United States and from countries around the world. The military officers, both international and from the United States, who attend the U.S. military's war and staff colleges are rigorously selected through standardized procedures by their home governments and likely to be promoted into the top echelons of their own militaries. Promotion statistics show that international graduates of U.S. war and staff colleges are likely to hold important jobs, and for some the highest rank, in their home country's armed forces. After graduation they belong to an international community of military officers who all understand U.S. military doctrine and U.S. perspectives on security issues. Additionally, the international graduates have had significant exposure to U.S. society. As part of their programs they have attended courses on U.S. government institutions, U.S. political processes, and U.S. perspectives on democracy and human rights. The international officers and their families have lived within the local communities. Their children have attended U.S. schools. Because these international officers are likely to be promoted into the top echelons of their own militaries, the ideas and perceptions that they learn while in the United States could have significant impact on their home countries' domestic and foreign policies. In the next two chapters, survey data is used to give voice to the many international students, so that we might learn more about their experiences while in the United States as well as the impact of the exchange experiences in how they think about the United States, its citizens, and its policies toward the rest of the world.

NOTES

1. The number of countries participating has remained fairly constant over the years. For example, in 2008, international participants came from 140 countries; in 2007, 134 countries were represented (DoD/DoS 2008).
2. As a comparison, the *New York Times* reported that the Global Hawk unmanned aerial vehicle costs $218 million per copy (Drew 2011).
3. In FY2010 the budget of the State Department was roughly $47 billion; the Defense Department's budget was roughly $553 billion.
4. To date only a very few women officers have participated.
5. The four military services are the U.S. Army, U.S. Marine Corps, U.S. Air Force, and the U.S. Navy.
6. The courses are also referred to by the terms "Intermediate Level Education (ILE)" for staff colleges and "Senior Level Education (SLE)" for war colleges.
7. As an example, roughly 20% of U.S. Air Force majors attend staff college in residence and roughly 15% of U.S. Air Force lieutenant colonels and colonels attend war college in residence.
8. These quotations are from the International Officer Survey that will be discussed in depth in chapters 5 and 6.
9. As discussed later in this chapter, the U.S. Army Command and General Staff College also welcomes a second smaller class that begins in February of each year.
10. For more information on the Field Studies Program, see chapter 3.
11. The following historical account of Henri LeComte's experience is drawn from Reichley (1994).
12. Today the course of study is only one year.
13. At this time it was known as the Army General Staff and the Army War College.
14. There were two officers from Pakistan and one officer each from Afghanistan, Algeria, Australia, Bangladesh, Bosnia and Herzegovina, Botswana, Brazil, Bulgaria, Burundi, Canada, Chile, Colombia, Croatia, the Czech Republic, Denmark, Egypt, El Salvador, Ethiopia, Germany, Honduras, India, Indonesia, Israel, Italy, Japan, Jordan, Kenya, Korea, Kuwait, Leba-

non, Lithuania, Macedonia, Malawi, Malaysia, Mali, Mexico, Mongolia, Montenegro, Morocco, Nepal, the Netherlands, Nigeria, Norway, the Philippines, Poland, Romania, Rwanda, Saudi Arabia, Senegal, Serbia, Slovenia, South Africa, Spain, Sri Lanka, Suriname, Sweden, Taiwan, Thailand, Turkey, Uganda, Ukraine, the United Arab Emirates, United Kingdom, Yemen, and Zambia (Cress 2011).

15. The history of the international programs at the U.S. Naval War College was drawn from the International Programs webpages at the Naval War College's website.

16. College of Naval Warfare (open only to U.S. military officers, senior-level), Naval Staff College (open only to U.S. military officers, mid-career level), Naval Command College (for international officers, senior-level), College of Naval Command and Staff (for international officers, mid-career level), College of Distance Education (a limited number of allied officers may attend), and College of Operational and Strategic Leadership (courses are for U.S. Navy officer and enlisted personnel)

17. Many thanks to John Charlton and Joycelyn Stevens of the National Defense University, who provided insight and information about NDU schools.

18. The National Defense Authorization Act for Fiscal Year 2012 (section 2861) redesignated the Industrial College of the Armed Forces as the Dwight D. Eisenhower School of National Security and Resource Strategy, hereafter referred to as the School of National Security and Resource Strategy.

19. The Goldwater-Nichols Department of Defense Reorganization Act of 1986 mandated numerous significant reforms that sought to force the military services to increase the interoperability of their equipment, organizations, personnel, and doctrine. The U.S. Congress mandated the reforms after several military debacles (such as the invasion of Grenada and the Iran hostage rescue attempt) demonstrated that individual service doctrines and parochial interests were interfering with the effective execution of military operations.

20. The officers obtained a position equivalent to the U.S. Chairman of the Joint Chiefs of Staff, U.S. Secretary of Defense, U.S. Secretary of State, U.S. Secretary of a military service, or achieved an equivalent rank as member of a multinational force or organization (NDU 2012).

5

Backgrounds, Social Integration, and Promotion Potential of the Exchange Officers

This chapter uses the model of socialization discussed earlier in the book in order to assess socialization in actual practice and from the view of the military officers who come to study at U.S. military war and staff colleges. The analyses are based on attitudinal surveys of foreign officers at U.S. military war and staff colleges. The chapter is divided into four sections and begins with a description of the survey, how and where the survey was conducted and basic demographic statistics of its participants. The second section assesses the level of international experience and previous cultural exposure of the exchange officers. The survey data show that the officers have considerable international experience and are certainly not "novices" when it comes to international politics. Unlike young high school or college students who go abroad on exchanges, the international officers have substantial ideological and professional ties to their home governments, they view their American experience as a tool to advance their careers, and thus they should be relatively impervious to socialization attempts.

The third section assesses the extent to which the international officers were welcomed and integrated into U.S. society and U.S. culture. It focuses on three aspects of the international officers' experiences: first, their level of social interaction with their fellow U.S. officers in the school environment and in their off-duty time; second, the extent to which being accompanied by their spouse and children facilitated social and cultural interactions that influenced their opinions of the United States and U.S. people; and third, the extent to which the officers obtained news and information from U.S. sources as opposed to non-U.S. sources. Social and professional integration is important to assess because the model of socialization presented in chapter 2 emphasized that the depth and extent of social interactions were important factors affecting the probability that new ideas and beliefs would be accepted. Additionally, it was argued that officers who were welcomed and integrated into U.S. society were more likely to take away positive impressions from their experiences in the United States than those officers who were not. This analysis complements the information presented in chapter 4 that discussed the many programs at the U.S. military schools that welcome, orient, and integrate the exchange participants and their families into U.S. culture. The surveys show that the officers felt they were accepted and integrated, both socially and professionally, into their U.S. military program, U.S. communities, and U.S. society. The officers felt that they were warmly welcomed and participated in many social events with their U.S. military counterparts. They and their families were embraced within U.S. communities, particularly through the sponsor programs. This experience then led to positive views of the United States and greater understanding of the values that the United States espouses.

The fourth section uses four measures to assess the promotion potential of the foreign officers who attend U.S. military war and staff colleges: first, the level of job they will have immediately after graduating from the U.S. military school; second, the number of people they will have under their command when they return home; third, the type of jobs that prior foreign graduates of U.S. military schools now have; and fourth, the officers' own thoughts on the impact the U.S. school would have on their own careers. The analysis of promotion potential presented in this chapter builds on information that was presented in chapter 4 detailing U.S. government data that tracks the success of the international officers and shows that a substantial portion of the officers go on to hold top positions in their military's command structure. Chapter 4 relied on data collected from U.S. sources whereas chapter 5 presents information from the international officers themselves. Both sources confirm that the international officers who graduate from U.S. military war and staff colleges are very likely to be promoted and occupy important positions in their home countries. Additionally, the data show that within many countries there are substantial cohorts of U.S. military war and staff college graduates who know each other. These officers have formed epistemic communities within their countries that maintain professional and personal contacts with a larger international epistemic community of fellow war and staff college graduates from the United States and other countries.

5.1. OVERVIEW OF THE SURVEY

In order to assess international officers' perceptions of the United States and evaluate the likelihood that these perceptions might matter in the future, international officers who were attending U.S. military war and staff colleges were surveyed. There is no office in the Department of Defense or Department of State that centrally administers the exchange programs for international military officers; rather, each military school runs its own program. Therefore, each war and staff college was contacted individually in order to obtain approval and conduct the survey.[1] At each school there were two sets of people who assisted with the survey: there were the operational staff members who run the exchange programs for the international officers and interact with them on a daily basis, and there were the headquarters staff members who oversee institutional research and work to ensure that research is conducted professionally and according to federal regulations.[2] At each of the schools the survey process was slightly different depending on the preferences of the schools. The survey was handed out in paper copies at the Army War College and Air University; at the other schools it was computerized and respondents were sent a link to the survey via an email message. But, otherwise, the surveys were identical.

The survey consisted of twenty-one questions. Some of the questions were multiple choice; others were open-ended, allowing the officer to write whatever he wished rather than be restricted to a limited number of responses. The survey gathered general background information, data on international experience, and data on career progression. There were several open-ended questions that asked general questions about the international officers' experiences while living in the United States as well as their perceptions about U.S. culture and U.S. people.

The timing of the survey was important. The goal of the survey was to understand the perceptions of the international officers after they had spent a considerable amount of time in the United States, so the surveys were distributed in the weeks immediately before the international officers graduated and subsequently returned to their home countries. In 2010, international officers at the National Defense University, Marine Corps University, Army War College, and the Army Command and General Staff College were surveyed. The same survey had also been given to the international officers at the Air Command and Staff College at Air University in 2006 and was also given to officers throughout Air University in 2011. The survey was voluntary for all respondents. All of the surveys that were returned were usable and the survey had an overall response rate of 61%.[3] Table 5.1 shows the number of respondents and the survey response rate for each of the schools.

Table 5.1. International Officer Survey Survey Response Rate

	Respondents	*Total (Response Rate)*
Air University	113	201 (56%)
Army Command and General Staff College	51	67 (76%)
Army War College	34	50 (68%)
Marine Corps University	19	29 (66%)
National Defense University	27	55 (49%)
	244	402 (61%)

5.2. HOME REGIONS AND GOVERNMENTS

The international officers come from a wide range of countries. The goal of the exchange programs is to have all regions of the world relatively evenly represented. There is usually only one officer representing any one country at any one school. This presented a problem in structuring the survey because asking the respondent to identify his home country was the equivalent of asking his name. While it would have been ideal to know the respondent's home country, this was not possible because of the need to ensure that each respondent should remain anonymous.[4] Instead, the respondent was asked to identify his home region. The regional distribution of the respondents' home countries is shown in table 5.2. The data show that the survey respondents came from across the world and they were relatively evenly distributed across the different regions. They represented a variety of cultural traditions with a significant proportion coming from non-Western cultural backgrounds. Only 17% of the respondents were from Western Europe, a statistic that helps to dispel the stereotype that most international officers who attend the elite military schools in the United States are from traditional U.S. allied nations in Western Europe.

Table 5.2. Home Regions of the Respondents

	Respondents	*Percentage*
Africa	30	12
Asia-Pacific	27	11
Asia, Central	4	2
Asia, South and East	30	12
Europe, Eastern	54	22
Europe, Western	41	17
Middle East	32	13
America, North, Central, or South	25	10
No answer	1	0
	244	99%

NOTE: Percentages do not equal 100% due to rounding.

This regional variation suggests that the respondents came from countries across the political spectrum; nevertheless, it was important for the analyses that follow to identify which officers came from nondemocratic countries. This was a challenging task because the respondents were not asked to identify their home country, so it was not possible to use country specific data from other sources in conjunction with the survey responses. Despite this limitation, the survey included two questions designed to measure the extent of democratic institutions and practices within the respondent's home country. First, the respondent was asked to rate how he thought U.S. citizens would rate the level of democracy in his country, and then, second, to rate how he saw his own country. In order to do this the respondent was given a scale with 0 representing "very authoritarian government" and 20 representing "very democratic government." This scale was specifically constructed to mirror the 21-point polity scale used to measure democracy-autocracy in the Polity IV Project that is one of the most widely cited regime ranking systems in political science (Marshall, Jaggers, and Gurr 2011). Any response that was 15 points or higher was labeled "Democratic" and any response that was 14 points or lower was labeled "Not Fully Democratic" for use in subsequent analyses.

As mentioned above, the respondent was first asked for his perception of how U.S. citizens viewed democratic governance in his country. The results for this question are shown in the top portion of table 5.3 labeled "Perceived U.S. View." In this assessment, a little over a third of respondents indicated that they came from countries that were not fully democratic. For the second measurement, the respondent was asked for his own view on the level of democracy in his own country. These results are shown in the next portion of table 5.3 labeled "Officer's View." In this assessment a quarter of the respondents rated their own country in the "Not Fully Democratic" range.

Because the respondents came from a large number of countries and from all regions of the world, the extent of democracy in their countries should more or less reflect the ratio of democratic to nondemocratic countries worldwide. The two results shown in the top half of table 5.3 suggest that respondents might have had some propensity to rate their country higher on the scale than an unaffiliated bystander would rate it. Table 5.4 shows the results from six difference of proportions tests that were used to assess the amount of bias in the respondents' democracy ratings and determine which of the respondents' ratings most closely

Table 5.3. Respondent Rating of the Democratic-Authoritarian Character of His Country

Perceived U.S. View		
Democratic	153	(63%)
Not Fully Democratic	85	(35%)
No Answer	6	(2%)
Officer's View		
Democratic	177	(73%)
Not Fully Democratic	62	(25%)
No Answer	5	(2%)
Polity IV Project, 2010		
Democratic	99	(60%)
Not Fully Democratic	65	(40%)
Economist's *Democracy Index, 2010*		
Full Democracy	26	(16%)
Flawed Democracy	53	(32%)
Hybrid Regimes	33	(20%)
Authoritarian Regimes	55	(33%)

Table 5.4. Proportion of Countries Identified as Authoritarian Difference of Proportions Test Results

	Z score	*Significant Difference*
Polity IV—*Economist*	1.32	no
Polity IV—Officer's View	3.21	yes
Polity IV—Perceived U.S. View	1.03	no
Economist—Officer's View	1.78	no
Economist—Perceived U.S. View	0.42	no
Officer's View—Perceived U.S. View	2.41	yes

NOTE: Two-tailed Z-test for difference in proportions at 0.05 significance level; values greater than 1.96 indicate that the proportions are significantly different

matched the state of democracy worldwide. The test result shown in the bottom row in table 5.4 (labeled "Officer's View—Perceived U.S. View") confirmed that the respondent was likely to have some bias in how he rated the level of democracy in his own country.

In order to determine which of the respondents' ratings most closely matched the state of democracy worldwide, the respondents' ratings were compared to two well-known rating indexes, the Polity IV Project's annual polity rating and *The Economist*'s Democracy Index. The bottom half of table 5.3 shows how these two rating indices assessed the level of democracy worldwide. The Polity IV Project tracks governance and publishes an annual assessment of democracy in countries worldwide. In 2010, the Polity IV Project rated 164 countries of which 40% were considered authoritarian based on the same scale that was used for the survey responses (a score of 14 points or below). Although widely used in political science, the polity index has not been without its critics who have pointed to its rather specific definition of democracy that excludes civil liberties of citizens.[5] The Polity IV index focuses on the nature of a country's political institutions, such as whether elections are competitive and the extent of constraints on the chief executive (Marshall et al., *Users' Manual* 2011).[6] The Democracy Index of *The Economist* provides another point of comparison. This index uses a wider concept of democracy encompassing measures of the political freedom of citizens, extent of political participation allowed, and respect for human rights by governments.[7] In 2010, *The Economist* assessed 167 governments of which 33% were rated as authoritarian.[8] As can be seen in table 5.4 (top row), Polity IV and *The Economist* were consistent in their ratings with no significant difference between the proportions of authoritarian countries each had identified. Table 5.4 also shows the results of the difference in proportions test for each of the respondents' two ratings paired with each of the democracy indices. The results indicate that the respondents' "Perceived U.S. View" aligns most closely with the two democracy indices. Thus, in the analyses that follow the respondents' "Perceived U.S. View" of the level of democracy in their home countries will be used to identify respondents from democratic countries and those from authoritarian countries.

5.3. INTERNATIONAL EXPERIENCE AND CULTURAL EXPOSURE

The survey asked several questions to assess how frequently the respondent had traveled abroad and, thus, was exposed to a variety of different cultures and peoples. The goal for this measurement was to assess the respondent's level of cultural and international exposure before he participated in his U.S. military exchange program. Previous research on socialization discussed in chapter 2 indicated that "novices" should be more amenable to socialization.[9] The military officers who had little previous international experience might be considered "novices" and it might be argued that they would be more amenable to their new experiences in the United States than those officers who had extensive international experience. Thus, those officers with extensive international experience should be less influenced by their experiences in the United States because they had already had many opportunities to form and reform their world views and would be resistant to new information. In order to assess the respondents' level of international sophistication, the survey used two measures: the number of languages that the officer spoke and the extent to which he had traveled outside of his own country.

The number of languages that a person speaks is one measure of cultural exposure and international experience. Table 5.5 shows the number of languages spoken fluently by the respondents.

Table 5.5. Number of Languages Spoken

Number of Languages	*Respondents*	*Percentage*
1	11	5
2	97	40
3	92	38
4	33	14
5	9	4
6	1	0
8	1	0
	244	101

NOTE: Includes native language; percentages do not equal to 100% due to rounding

At a minimum all of the officers were able to communicate in English (a requirement for attendance in the war and staff college programs) and in their native language. As can be seen in table 5.5, most of the officers (95%) spoke at least one language in addition to their native language, and over 50% of the officers were fluent in three or more languages. On average, the officers who came from nondemocratic countries spoke more languages than those who came from democratic countries.[10] Of the eleven officers who spoke five or more languages, six of them (55%) came

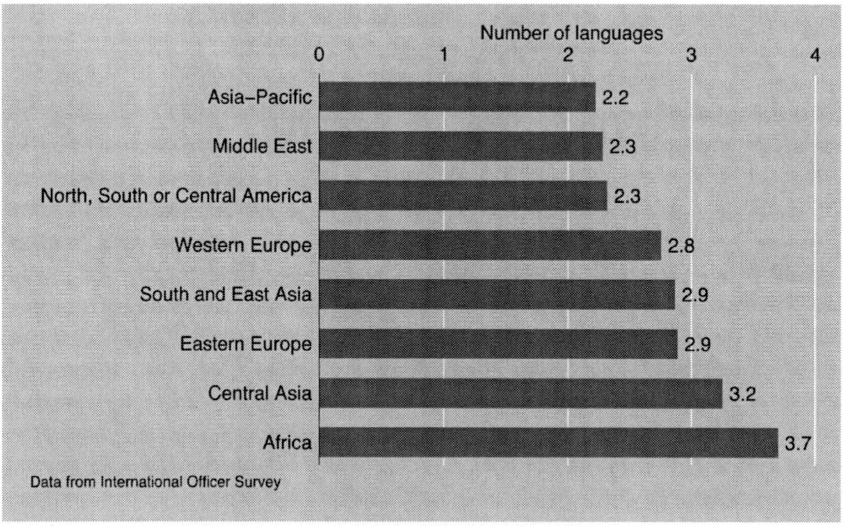

Figure 5.1. Average Number of Languages Spoken by Region

from Africa. The officers who spoke only one language were predominantly from Western Europe (36%) and Asia-Pacific (36%). These regional variations are reflected in figure 5.1, which shows the mean number of languages spoken by region.

The officers with more extensive experience operating in multilingual settings came from nondemocratic countries and from developing regions of the world (Africa, Central Asia, and Eastern Europe).

The extent to which the respondent had traveled outside of his home country is another measure of previous international experience as well as exposure to different cultures. Figure 5.2 shows the number of countries besides the United States that the respondents had visited in the previous five years.

Most of the officers had visited several other countries; only 21 (9%) had not traveled outside of their own country. Almost half of the officers (48%) had traveled to five or more countries. The European officers were by far the ones with the greatest breadth of travel. Of the thirty-three officers who had traveled to ten or more countries, 42% were Western Europeans and another 34% were Eastern Europeans. The officers who had not traveled abroad were fairly evenly spread across all of the other non-European regions. This variation is shown in figure 5.3, which shows the average number of countries that an officer had visited broken down by the officers' home regions. On average, officers from democratic countries traveled to 1–2 more countries than officers from nondemocratic countries.[11]

In terms of international experience and cultural exposure, the officers overall had a significant amount of previous experience. In general, the majority of the international officers speak at least three languages. This certainly indicates a level of cultural sophistication in terms of their ability to communicate across cultural-linguistic groups. In terms of travel, over 90% had traveled outside of their home

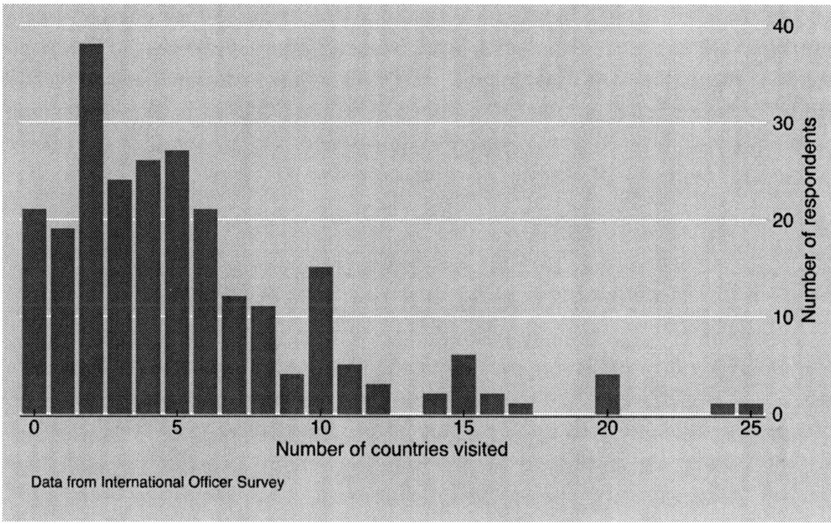

Figure 5.2. Number of Countries Visited in the Last Five Years

country in the last five years. On average, those officers who had traveled visited six other countries or took more than one trip abroad each year, in addition to their current time in the United States.[12] Overall, the officers had a good deal of previous international experience; they were certainly not novices with respect to foreign languages and foreign cultures. In this sense, the international officers may be more resistant to new ideas and new ways of doing things than, for example, college students

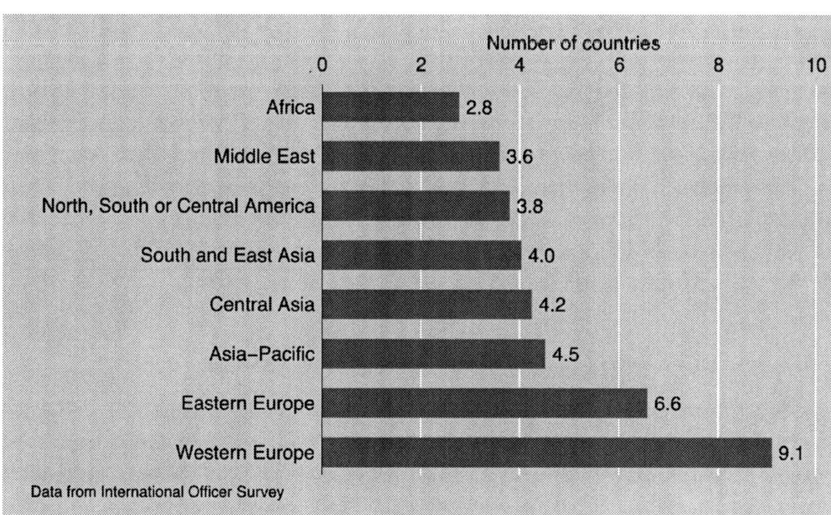

Figure 5.3. Average Number of Countries Visited by Region

who go abroad to study. The officers have also been selected by their governments to represent their countries and, in that sense, certainly come to the United States well versed in the military and foreign policies of their home countries. These officers, it should be remembered, are the rising elite in their home countries; they were chosen because they are loyal to their home government; and they are certainly not novices in terms of international politics and cross-cultural experiences.

5.4. INTEGRATION INTO U.S. CULTURE AND SOCIAL ACTIVITIES

The model put forward in this book emphasizes the importance of social interaction as a key to socialization. Research has shown that international exchange programs that provide students with opportunities for extensive interactions with people in the host country are much more likely to result in positive attitudes toward the host country (Atkinson 2010, Miller 2006, Selltiz et al. 1963, Sunal and Sunal 1991, and Ye 2001). As discussed in chapters 3 and 4, one of the key characteristics of the exchange programs hosted by the U.S. war and staff colleges is the very extensive integration of the international officers with their U.S. counterparts in the classroom and with a broad range of U.S. people in various social activities outside of the classroom. There are many ways the officers are integrated, from their classroom studies to sponsor programs to children's activities. This is certainly true for the officers who attend the war and staff colleges run by the U.S. Army, the U.S. Air Force, the U.S. Marine Corps, and the National Defense University. In these schools, the exchange officers sit side-by-side day-after-day with their U.S. counterparts. Because of the way the classrooms are organized by the schools, it is quite clear that during their on-duty time the international officers are socially and professionally integrated with their U.S. counterparts. But to what extent do the exchange officers spend their off-duty time with U.S. people? The survey contained three measures of social integration into U.S. society. From their experiences in the United States, the international officers were asked for their assessment of their off-duty activities, their assessment of their families' involvement, and their sources of news on world events. These measurements are discussed in the three subsections below. The data show that most of the international officers spent a good deal of time interacting with U.S. people during their off-duty hours (as well as during their on-duty time).

5.4.1. Interaction with U.S. and Other International Officers

While the exchange officers participated with their U.S. counterparts in the activities in their schools, it might be the case that they had few social interactions with other U.S. people outside of school. While the educational environment is certainly key to learning and socialization, social contacts and social interactions outside of the school environment are important as well. In order to assess the extent to which an

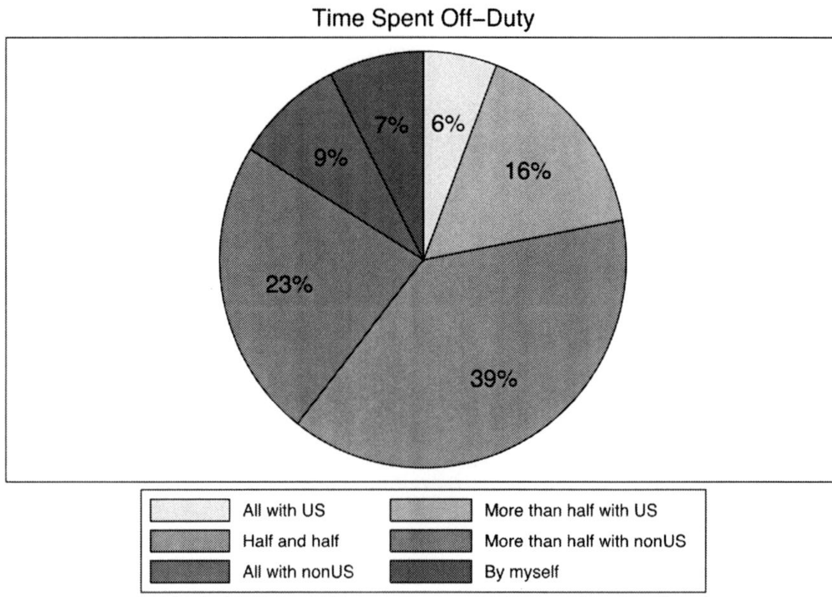

Data from International Officer Survey

Figure 5.4. Interaction with U.S. and Other International Officers during Off-Duty Time

international officer spent off-duty time with his U.S. colleagues and U.S. friends, the officer was asked to assess the amount of time he spent with U.S people in contrast to interactions with other international officers or non-U.S. friends. It could also be the case that the officer spent most of his off-duty time by himself—this, obviously, would limit the number and type of social interactions in which he participated. Figure 5.4 shows the amount of off-duty time the respondents indicated that they spent within each of these three different categories of people.

These results show that the international officers spent a good deal of their off-duty time with their U.S. colleagues and other U.S. citizens. Overall, 61% of the officers spent at least half of their off-duty time with U.S. people as compared to 32% that spent most of their off-duty time with their fellow international officers. In considering cross-cultural interactions, 93% of the officers indicated that they spent their off-duty time with either their fellow exchange officers (and their families) or with U.S. people. Only 7% of the officers indicated that they spent most of their off-duty time by themselves. These eighteen officers were evenly spread across all regions except Asia-Pacific, which only had one respondent in this category.[13] Figure 5.5 shows the same data broken down by whether the officer came from a democratic or nondemocratic country.

There is a significant difference[14] between officers who came from democratic countries as opposed to those from nondemocratic countries with 66% of officers

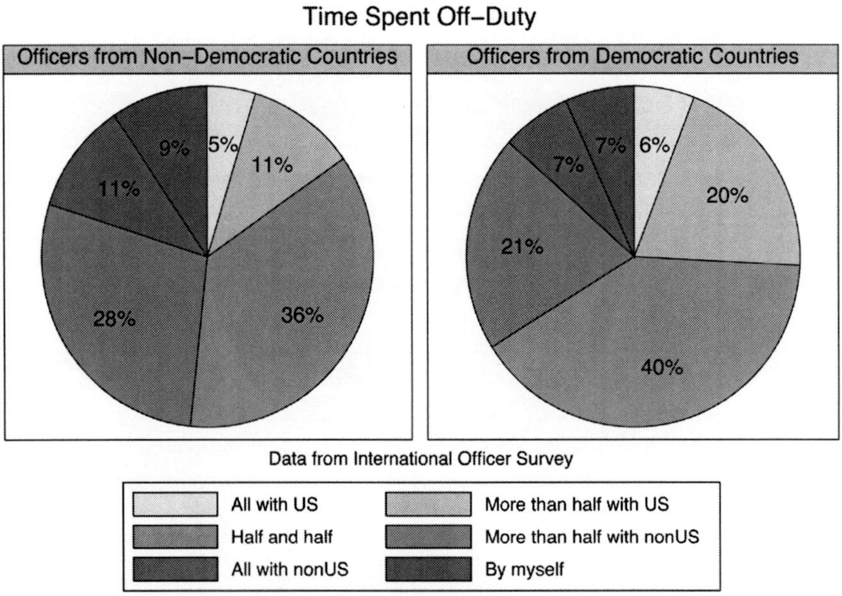

Figure 5.5. Interaction with U.S. and Other International Officers during Off-Duty Time by Regime Type

from democratic countries and 52% of the officers from nondemocratic countries indicating they spent at least half their time with U.S. citizens. This is an important difference to notice, but it is also noteworthy that more than half of the international officers from nondemocratic countries spent at least half of their off-duty time with U.S. people. The above data, in combination with the fact that all of the international officers' on-duty time was spent in the classroom with their U.S. counterparts, indicates that, in general, the international officers spent a significant amount of their time interacting with U.S. military officers and U.S. people in their local communities. Many of the comments that the officers made on the survey also support this conclusion. For example, in the comment section at the end of the survey an officer from the Middle East wrote that his year as an exchange student "was a great opportunity to interact with US military officers of different diversity. I learnt a lot from them as they learned from me. I have met and established great friends. The friendships will last for a long time."[15]

5.4.2. Importance of the Family Experience

For the international officers, one significant benefit of their exchange programs is the ability to bring their families with them to the United States. All of the schools welcome families. They have programs to assist families in settling into life in the

United States, and the schools hold many events geared toward involving the spouses and children in social activities, from parties to volunteer work. In recent years, the schools have organized a number of programs for spouses including English language classes. The ability of the international officer and his family to navigate U.S. society and his interaction with fellow officers, faculty members, sponsors, and other people that they meet in the United States certainly play a very important role in the attitude that the international officer forms toward the United States, the U.S. military, and U.S. people. Overwhelmingly, when asked about the impact of the experience of living in the United States on their families, the international officers were very positive and very enthusiastic. Representing the sentiments expressed by many of the officers, an officer from South and East Asia wrote: "The experience here will probably be unforgettable and valuable even after to my country. I believe it will influence my children for the rest of their lives, especially." An officer from Eastern Europe wrote that his yearlong stay in the United States was a "Very good experience for my family. It will help them for the rest of their lives."

Families are an underevaluated aspect of the exchange officers' overall experience. The survey found that families accompanied 84% of the respondents.[16] Figure 5.6 shows that there is significant regional variation in the extent to which families accompanied the military officers. The regions of the world where the United States might most like to build partnerships, Central Asia, Africa, and the Middle East, are also the regions that have lower levels of family participation. The same is true for the types of countries that the officers came from. Only 79% of the officers who came from nondemocratic countries were accompanied by their families as opposed to 86% of the officers from democratic countries.

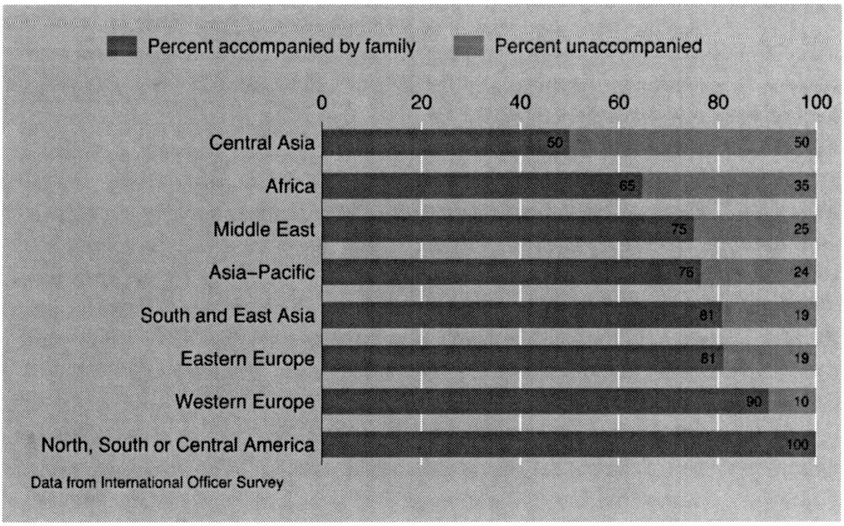

Figure 5.6. Percentage of Officers Accompanied by Their Families by Region

Table 5.6. Impact of the Exchange Experience for the Officers' Families

Impact	*Percentage of Respondents Citing This Item*
Learned or improved English	38
Increased cultural awareness	32
Gained greater appreciation for the United States	24
Broadened perspective of the world and people	21
Gained more knowledge of the United States	17
Made many good friends	15
Children got good education at U.S. schools	9
Wife became actively involved	6
Children and spouse now have increased confidence	5
Gained better understanding of U.S. democracy	4
Children now have more opportunities in their home country	4

NOTE: Some respondents described the impact using adjectives; some described specific ways that the experience had affected their families. This is the percentage of times an item was mentioned by those officers who described at least one specific impact. Because some officers described two or more impacts, the percentages do not equal 100%.

In reading over the survey responses from the international officers, it is quite clear that the family experience was very important to them. The response to the family programs was overwhelmingly positive with every single respondent recommending that officers should bring their families with them. The international officers enumerated many ways in which the family experience was beneficial to themselves and all of their family members. Table 5.6 lists the impacts that the respondents mentioned.

The most-often-cited impacts involved opportunities to learn English, to live in a foreign culture, and to build friendships with people from around the world that resulted in a greater appreciation of other cultures and a broader perspective on the lives of people around the world. For example, an officer from the Americas[17] said that the experience was "Very positive for my family. My wife became actively involved with the conversation + culture program. My children attended a very good high school. We travelled throughout the US. Overall, a very positive experience."

The single most frequently cited benefit was the ability for the entire family to learn and improve their English language skills. An officer from Africa wrote: "My family has learned English, especially my son who became an English speaker (besides French and Arabic). My family also became more familiar with American society (concerning habits/culture . . .)." An officer from Eastern Europe stated that for his family there would "be a great impact. Now, they're very good or, in fact, excellent English speakers; they really enjoyed contact with the American culture and education system, as well as the contact with other cultures." The opportunity for the officers' children to learn English is one benefit that is cited by a very large number of the international officers. The officers have noted that their children learned to speak

English rapidly, developed many friendships with other children, and participated in a wide range of local community activities for children. As an officer from Western Europe remarked, "My three-years-old daughter is now speaking better English than I will ever be able to."

In their survey responses many of the officers also cited the ability to experience a different culture and the opportunity to build cross-cultural friendships as very important experiences. For example, an officer from Eastern Europe stated the impact on his family: "First of all, they improved their knowledge of English. Furthermore, they became more open towards other people, and got accustomed to cultural, and regional differences." A senior officer from the Middle East stated, "They like the country and made new American friends." The officers noted that their families were able to experience different cultures, formed friendships that they hoped would last for a long time, and in the process gained important insights into U.S. people and U.S. society. As an officer from South and East Asia remarked, "The main impact is the ability to speak English, it will be getting better and more fluent. The culture and way of life in America is also influencing them, especially in the openness, fairness and law enforcement." Another officer from the same region (different school) expressed a similar sentiment: "It is a very good eye opener and has definitely widen their horizon in many aspects. They now have a better and positive perspective of the U.S."

For the officers who bring their families with them, family travels, opportunities for their children, and interacting with other families are often cited as having a very positive impact on their attitudes toward and perceptions of the United States and its citizens. One respondent from the Middle East remarked, "I made a trip to Disney world in Florida by my car. The experience was exciting because we saw beautiful places and explored the beauty of America." And from this and his other experiences he observed that "the media and the policy do not reflect the good nature of the American people." An officer from South and East Asia had a similar comment. He wrote that his family had "good memories of the U.S. We met many people, made new friends and understand the people of the U.S. better, especially in [the local area]. Having been here for a year, my family and I fully agree that there are many more nice people in U.S. other than those portrayed in movies especially those movies with a lot of violence in them."

Family activities and experiences of the officers' wives and children contributed greatly to the officers' positive views of the United States and ordinary U.S. citizens. These experiences played an important role in overcoming the negative portrayal some officers had seen of the United States in the media and negative opinions of U.S. foreign policy, as the two comments above illustrate. The officer's perceptions and attitudes are also a function of the type of experience that his entire family encounters, from trips across America to children's sports and educational opportunities. The children of the international officer are an underestimated, yet very important, element in the socialization of the officer because the children pull the entire family into many types of fun social and community activities. The officers, as parents, may also experience the world through their children's eyes.[18] For example, an officer from South and East Asia, wrote, "I have 2 sons (13 years old and 11 years old). I think that they had a very good experience. Both of them joined in a basketball team in school and YMCA. They had a lot of friends and my son's ability of English improved very well." An officer from the Middle East observed his family's experience: "It was a nice opportunity to learn about American culture. We have formed a positive opinion about Americans. My children studied and learned nice things that will benefit them when they return [home]. My wife expressed great interest in volunteering."

The role played by the officer's children and spouse are an underappreciated but significant factor in the quality of the exchange officer's experience. Some of the most memorable experiences for the officers are the social interactions that they have with their U.S. counterparts and their families, with their sponsor families, with other international officers and their families, and with the people the officers meet through their children's activities. For example, family trips to Disney World in Orlando, Florida, were specifically cited by twenty respondents as one of their best experiences while in the United States. As one respondent from South and East Asia remarked, his best experience while in the United States was "the Disneyworld visit. [It was] a dream come true for my daughters. In fact, the whole year is a great adventure." Another respondent from the Middle East remarked that the best experience for him and his family was "visiting our friends' homes (Americans and others) and inviting them to visit us."

There are some international officers who did not bring their families with them primarily due to the high cost of living in the United States, but also due to dual careers and other considerations such as the affordability of health insurance. As mentioned earlier, these unaccompanied officers are more likely to come from authoritarian countries rather than democratic countries; and as figure 5.6 shows, they are more likely to come from regions of the world where U.S. military-to-military partnerships are less institutionalized. This is unfortunate because the family experience is one of the most significant factors impacting the opinions that the exchange officer forms of the United States and U.S. people. This is one of the most important findings of the research presented in this book: the family is a "force multiplier" when it comes to the socialization of the exchange officer.

In the survey, several of the officers who were unable to bring their families with them took the time to remark on the positive experience that they had observed for their fellow international officers who were accompanied. For example, an officer from the Middle East who was only able to bring his family to the United States for a short visit stated that it would have been great to have his family with him for a longer time, but even during his family's short visit the experience was memorable: "They came with me for a short time. They asked me to buy a yard for them because we live in an apartment!!!!!. I think the wives and kids enjoy a lot of care and opportunities with all available facilities and programs on/off post." An officer from Africa stated, "I think it would have been a good experience" had he been able to bring his family. Another officer from Africa whose family was with him for only the last month of his schooling stated that there was not much impact on his family because their stay was so short, but for other families "it's just an experience of a lifetime!" Echoing that sentiment was an unaccompanied officer from South and East Asia who thought the impact on his colleagues' families was "significantly important, but unfortunately my family, especially my children couldn't have this chance."

The importance of the exchange experience for the whole family should not be underestimated or underfunded. For an incredibly small amount of grant funding to help officers bring their families with them, the positive family experience could be enjoyed by a greater number of officers, many of whom come from democratically austere regions of the world. Funding family travel, health care, and housing for those international officers who cannot afford to bring their families with them could generate a tremendous amount of goodwill toward the United States. This is a matter beyond the control of either the U.S. Department of Defense or the U.S. Department of State who are limited by the Arms Export Control Act from using their resources to fund family travel and family health-care insurance. It is an issue that can only be resolved through Congressional action.

5.4.3. News Media as Sources of Information

Person-to-person interactions are certainly a key avenue through which information is exchanged and perceptions are formed. Another method of acquiring information is through various news sources from television to the Internet or by just talking with other people. The comments of the international officers indicate that some of them came to the United States with negative perceptions of U.S. people and U.S. society because of what they and their families had seen in news media and in U.S. movies. The personal experiences they had while in the United States seemed to conflict with the views that they had previously formed from media sources. Thus, it is interesting to explore the sources of their information.

The officers were asked to identify where they get most of their news about international events. Figure 5.7 shows that the vast majority of the officers got their international news from the Internet, an international source of news. This was the case across all regions. U.S. television was also cited by a significant number of respondents, and it is interesting to note that over 16% said that talking with other people was an important source of international news for them. As the officers themselves noted it seems that many of the stereotypes that they have of U.S. people and U.S. society come primarily from U.S. media. For each source of news where the origin was specified (for example, U.S. newspapers versus non-U.S. newspapers), in every case more respondents relied on the U.S. version rather than the non-U.S. version.

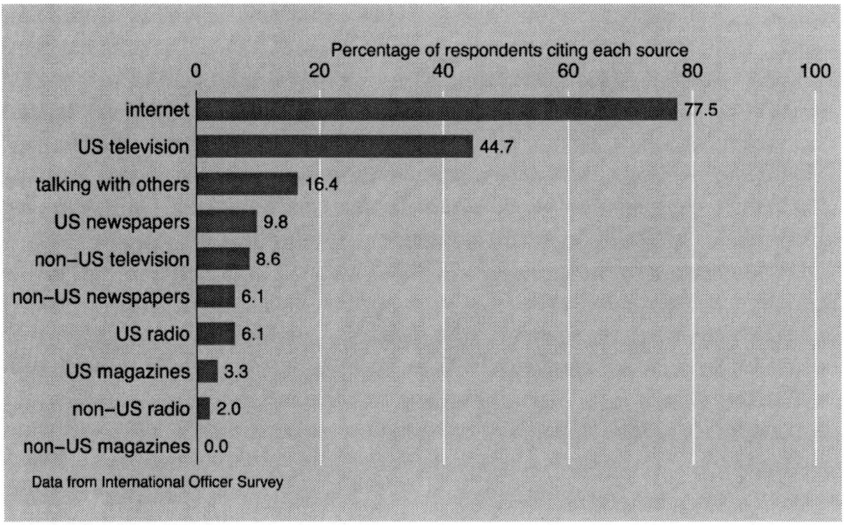

Figure 5.7. Primary Sources of News

5.5. FUTURE CAREER PROSPECTS

The likelihood that an exchange officer will be able to shape future policies in his home country depends, of course, on the officer's ability to achieve command positions where he can influence subordinates, peers, and policymakers. As mentioned earlier, a considerable number of graduates from all of the war and staff colleges have achieved the highest positions within their armed forces, and a few have become heads of state. But what are the future career prospects in general for all of the officers?

Overwhelmingly, the officers themselves believe that graduation from a U.S. war or staff college will have a very significant and positive impact on their careers. One mid-level officer from Eastern Europe stated that both his immediate and long-term career prospects were greatly improved by graduating from one of the staff colleges. He wrote, "[The staff college] is very famous and high-ranking institution and after coming back home every student from my country receives very good position and usually has very good military career." Another officer from Western Europe summed up the impact of the school on his future career with one capitalized word: "HUGE." Other officers echoed this sentiment and used adjectives and expressions such as "tremendous," "enormous," "great," "very big," "definitely positive," and "undeniably positive" to describe the impact on their future military careers. Of the officers who commented, 98% thought it would help their careers, while only 2% thought it would have little impact.[19]

The survey asked several questions designed to assess the career prospects for all of the international officers who come to the war and staff colleges in the United

States. First, the officers were asked the level of job that they would return to after graduating from the war or staff college. Second, the officers were asked how many people they would supervise in their next job. Third, the officers were asked if they knew of other people in their own militaries who had graduated from a U.S. war or staff college, and, if so, what level of job these people currently had. Finally, the officers were asked what impact they thought their attendance at the U.S. military war or staff college would have on their future careers. Each of these measurements is discussed below. To summarize, the results from these questions confirm what the U.S. government data showed: the exchange officers are indeed very likely to hold future positions of influence in their countries.

5.5.1. The International Officer's Next Job

The job that is assigned to the officer after his graduation from the U.S. military war and staff college is one indicator of his future success. If the officer were assigned a job with more responsibility than he had before coming to the United States, this would indicate that his home country values the knowledge and experience the officer gained while in the United States. The survey data show that 73% of the international officers were going home to jobs of increased responsibility and only 20% were returning to a job that was similar to the one they had before.[20] Figure 5.8 provides a breakdown of the data for officers returning to nondemocratic countries as opposed to those returning to democratic countries. As can be seen, there was

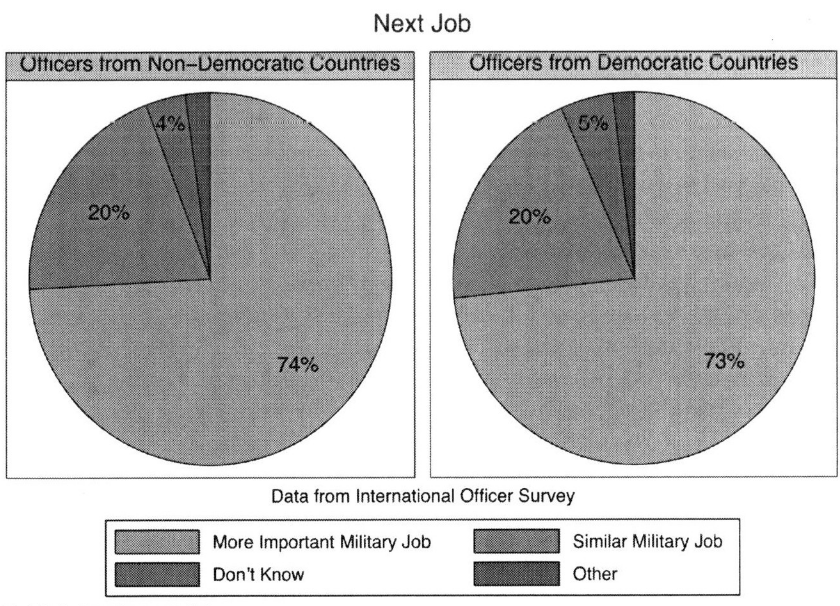

Figure 5.8. Level of Job after Returning Home by Regime

essentially no difference between these two categories with most of the officers from both types of governing systems returning to military positions with more responsibility than the position they had before coming to the United States.

5.5.2. Command Responsibility

Another indicator of future career prospects for the officer is the number of people that the officer would command or supervise in his new job. For example, four of the officers who had indicated they were going back to a job similar to the one they had before coming to the United States also indicated that in this job they would be supervising more than 1,000 subordinates, and one of these officers was going to supervise over 10,000 people. Thus, the job that they were returning to was already a high-ranking command position. The number of people supervised is a rough approximation of future career potential because some officers may go on to command positions where they would supervise thousands of soldiers, while others might go on to headquarters staff positions where they might wield tremendous power but directly supervise only a few people. Additionally, some countries have large militaries and others have smaller militaries, which impacts the potential number of people an officer might supervise. Despite all these caveats that are good to keep in mind, this measure does provide another layer of information about the officer's immediate future.

It is useful to have an idea of the command responsibilities that an officer might take on after he has returned home, for at least two reasons. First, the number of subordinates that an officer commands is an indication of his future promotion potential. Second, a commanding officer provides his vision of appropriate behavior to his subordinates and, within the confines of his country's military regulations, enforces that vision. In other words, the commanding officer is responsible for the military socialization of his subordinates. This is one way that new ideas can be introduced within military organizations. Certainly, this could also be a task that has been assigned to the officer by his government. The officer was specially selected and sent for a year of schooling in the United States for just this very reason: to learn new ideas and new ways of doing things. One of the tasks of the returning officer may be to improve his own military's effectiveness by introducing new ideas in the form of new operating procedures, new organizational concepts, or new doctrine that he learned while in the United States. This is not to say, however, that these are the only types of ideas that might be introduced. For example, if the commanding officer has a positive view of the United States and the U.S. military this may be something that he communicates to subordinates through either his official functions or in informal interactions that he might have with them. Table 5.7 shows the number of people the respondent indicated that he would command or supervise when he returned home. At the time of the survey, many of the officers did not know how many people they would supervise, but for the officers who did know, many were returning to substantial commands. As can be seen, 100 officers (57%) indicated that they would have 100 or more subordinates, thirty-five officers (20%) would command 1,000 or more subordinates, and three officers returned home to supervise 10,000 or more people.

Table 5.7. Number of People under the Officer's Command

Number of Subordinates	*Respondents*	*Percentage*
0	11	6
1–9	17	10
10–49	25	14
50–99	21	12
100–499	41	23
500–999	19	11
1000–4999	26	15
5000–9999	6	3
10,000 or more	3	2
"Many"	6	3
	175	99%

NOTE: Percentages are based on 175 responses. Forty-seven officers indicated that they did not yet know how many subordinates they would supervise; twenty-two officers did not answer (most likely because they also did not yet know). Six officers wrote "many," which was important information but impossible to quantify, so "many" was included as a separate response category. Percentages do not equal 100% due to rounding.

Figure 5.9 and figure 5.10 show the regional variation in command responsibilities for the new international graduates of U.S. military war and staff colleges. Figure 5.9 shows the average number of subordinates by region and figure 5.10 shows the number of officers from each region that will have 1,000 or more people under their command. As can be seen, Middle Eastern officers were much more likely to command large units than their international counterparts from other regions.

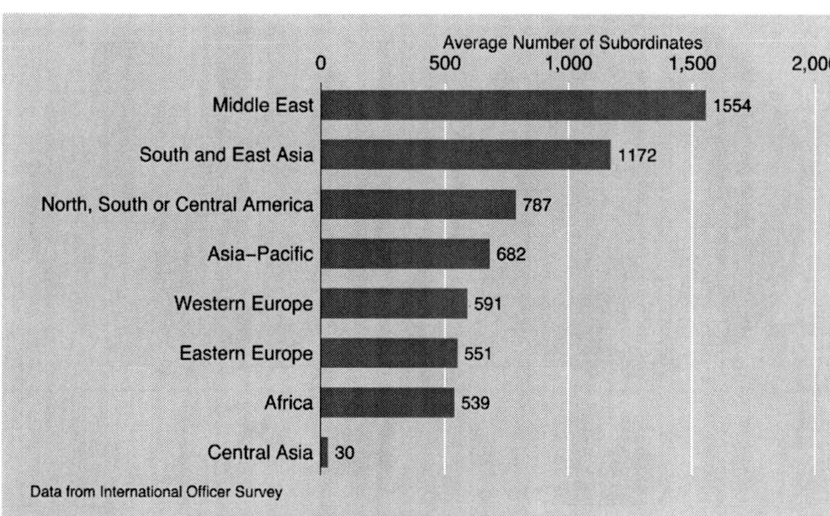

Figure 5.9. Average Number of Subordinates by Region

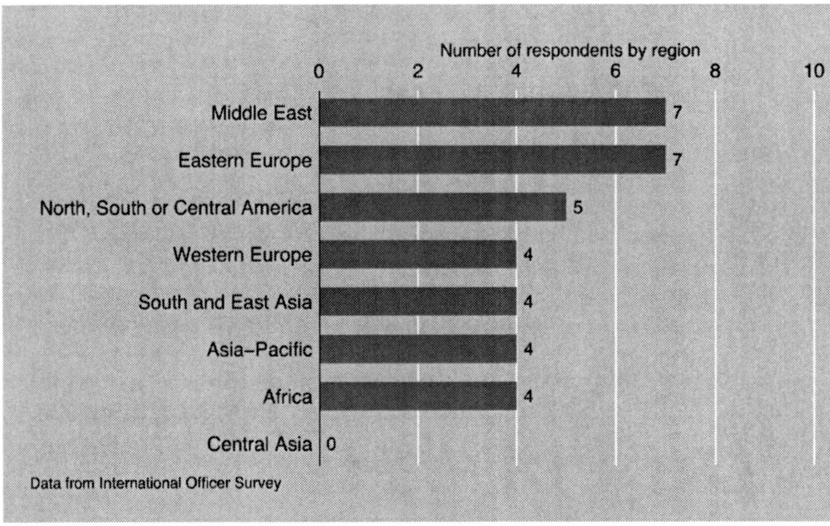

Figure 5.10. Number of Officers Who Will Supervise 1,000 or More People by Region

Figure 5.11 shows the breakdown in command responsibilities for officers from nondemocratic countries as opposed to those from democratic countries. There is very little difference in the size of command responsibilities between these two groups of officers.[21]

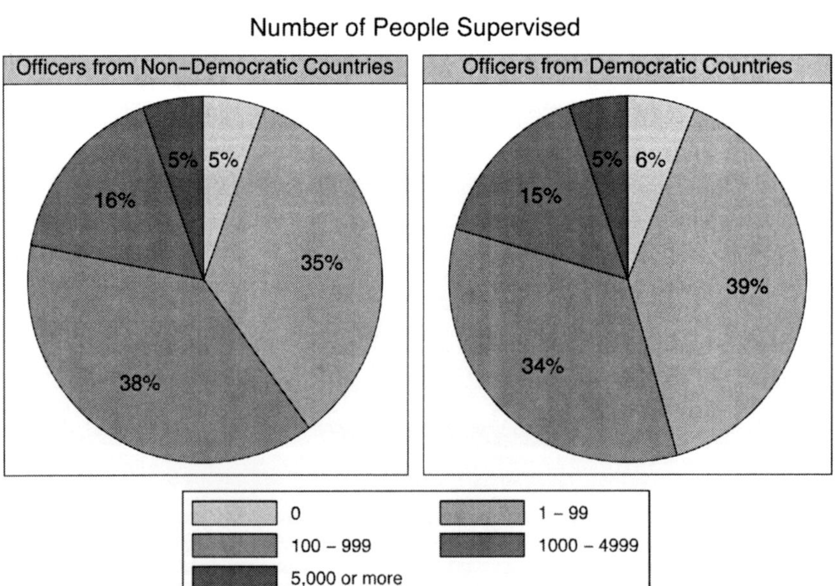

Data from International Officer Survey

Figure 5.11. Number of People Supervised by Regime Type

5.5.3 Current Jobs of Previous Graduates

The accomplishments of previous graduates of the war and staff colleges are another indicator of the future career prospects for those who are still in school. The halls of fame for the international officers at the war and staff colleges have recognized hundreds of graduates who have gone on to the highest ranks in their own countries. The hall of fame inductees occupy (or formerly occupied) positions of considerable power and influence. At least a few of the officers that were surveyed for this book will undoubtedly also go on to achieve the highest ranks in their military or government. This expectation was, in fact, expressed by several of the respondents. For example, when he was asked what impact the U.S. military school attendance would have on his career, one officer remarked "Hope it will have the same impact as it had with those international officers inducted in the [college's] hall of fame." This might indeed be an achievable goal.

In order to get an idea of the career progressions of graduates in general, not just those who have achieved these highest level positions, the survey asked the officers about other previous graduates of U.S. military war and staff colleges. The respondent was asked if he knew a previous graduate and, if so, to indicate the level of responsibility of military or civilian jobs that this person now had. Over 97% of the respondents indicated that they knew someone from their home country who had previously graduated from a U.S. military war or staff college.[22] There were seven officers who indicated that they knew of previous graduates in both very important military positions and in very important civilian positions. A senior-level officer from South and East Asia wrote that most previous graduates from his country had reached very influential positions: "It has a very big impact on my future military career. We have more than 30 officers who graduated from [the war college], and most of them have already been promoted to General." The fact that so many of the international officers knew of previous graduates is important because it indicates that within the majority of countries that participate in the exchanges, there is a cohort of U.S.-trained military officers who are familiar with each other and who have shared similar experiences in the United States. As discussed earlier, an epistemic community can be a key support mechanism for "idea entrepreneurs" from within the community as these members seek to promulgate new ideas within a larger setting. Not only were previous graduates well known, but also the majority of them had gone on to hold "very important" positions. As can be seen in figure 5.12, 67% of the respondents indicated that they knew previous graduates who now held "very important" military jobs.

Table 5.8 provides the breakdown by region and shows that there were a substantial number of respondents from each region who knew of a previous graduate who now held a very important position.

Figure 5.13 shows the breakdown in responses by type of governing system. Respondents from nondemocratic countries knew more U.S. military war and staff college graduates who now occupied "very important" civilian positions than the officers from democratic countries. For this survey item, this was the only statistically significant difference between respondents from nondemocratic countries and those from democratic countries.

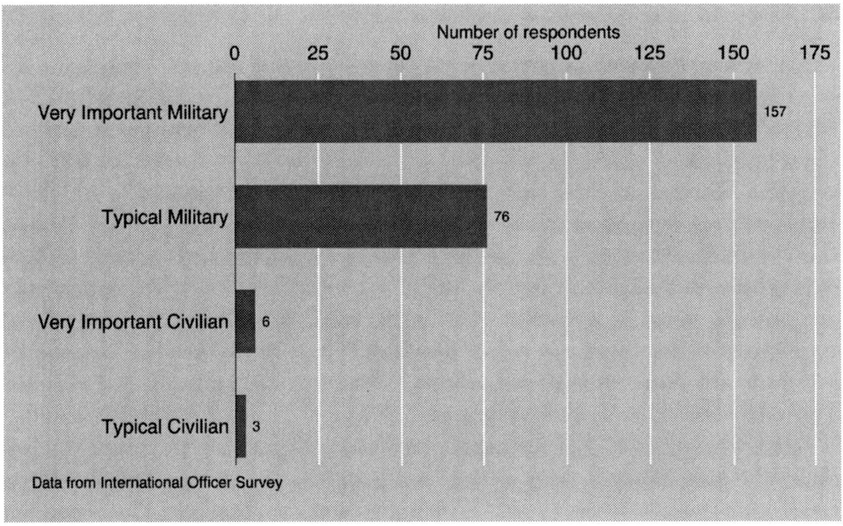

Figure 5.12. Jobs of Previous Graduates

5.5.4. Officers' Personal Evaluations

Finally, another way to assess the impact of their U.S. military school experience is to ask the officers themselves. For this purpose, the survey contained the open-ended question: "What impact do you think that your attendance at [name of school] will have on your future military career?" Open-ended questions such as this are very informative because the respondent is not restricted to any set of choices but can write whatever he wants to write. The responses to this survey question were categorized based on the answers that the respondents wrote. In most cases, an officer listed or described more than one impact. Table 5.9 shows the various impacts mentioned and the number of times each was mentioned.

Table 5.8. Respondents Who Knew a Previous Graduate in a Very Important Position

Region	*Percent of Respondents*
North, South, and Central America	76%
Africa	73%
Middle East	72%
Eastern Europe	70%
South and East Asia	67%
Asia-Pacific	63%
Western Europe	54%
Central Asia	50%

NOTE: These are the percentages of respondents from each region who knew of a previous graduate in a very important civilian or very important military job.

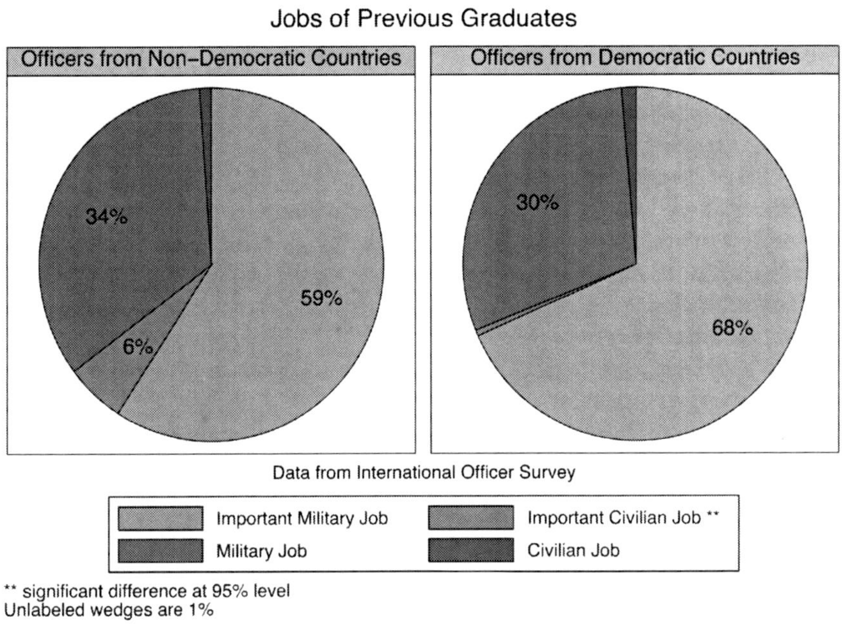

Figure 5.13. Jobs of Previous Graduates by Regime Type

Table 5.9. Impact on Future Military Career

	Number of Times Cited	*% of Respondents Citing This Item*
Greater promotion and career opportunities	93	56
Developed personal skills and insights	46	28
Gained a wider perspective	24	14
Became a better leader	18	11
Acquired knowledge of U.S. military and U.S. government	12	7
Increased opportunities to work in a multinational setting	11	7
Developed skills and knowledge to help own country	11	7
Gained increased ability to work with the U.S. military	7	4
Gained a network of friends and professional contacts	7	4
Became more self-confident	5	3
Acquired better understanding of U.S. culture	3	2
Ceremonial	1	1

NOTE: Percentage does not equal 100% because many respondents cited more than one impact. This question was asked of 167 international officers; the question was not asked on the ACSC survey.

Roughly half of the respondents specifically mentioned that attendance at the school would increase their chances of promotion and had opened up new opportunities for them. It is the same for officers from nondemocratic countries and those from democratic countries with no statistical difference between these two groups. This sentiment is reflected across the different regions. An officer from Asia-Pacific wrote that the school would have a "tremendous impact" on his career. He noted many benefits of the exchange experience: "I surely gained knowledge, group manners, negotiation skills and resiliency during my course. Surely that will serve as a huge boost for my career." An officer from the Americas made a similar assessment for his future: "This will make me competitive for command positions. I have a much better chance to make general officer." Another officer from Asia-Pacific remarked, "It has been a positive experience that has taught me a great deal. Historically, graduates of this institution generally go on to have very successful careers within our military."

In addition to achieving high rank, many of the officers mentioned specific benefits and their plans to use their experience after returning home. A senior-level officer from Africa hoped to use his experience to institute changes in his own country: "The impact of the [war college] on my future military career will be enormously positive. I hope I would be able to bring some changes." Another officer from Africa detailed how the program had changed his own perspectives. He wrote that his time in the United States had "helped me to begin to think more strategically about my service and its impact on national security. It has also sharpened my history and broadened my perspective on international relations including why and how our service culture became what it is. This has given me an understanding of the patience required to bring lasting change and the networking and HR [human resources] development I will have to foster in order to build a legacy as I rise in rank and responsibility." A senior-level officer from South and East Asia cited both professional and personal impacts and the desire to use his experience to bring improvements to his country: "It was one of the most important experiences and periods in my military career, of course, it would have a significant impact, not just in my military, rather, in my civilian life as well. I hope by having good experience from my presence in US, I can do my best for my country and people." A senior-level officer from Eastern Europe also cited plans to use his experience: "I would say great impact. Being at the [war college] for a year, reading materials about strategic issues, listening to my colleagues, from the U.S. and other countries, opinions about different strategic issues, was a great experience for me. And in my next assignment, I will use a number of lessons learned and gained here."

Officers also cited specific impacts such as improved English, having a wider perspective, and thinking of issues at the strategic level as useful skills that would help them to advance. An officer from Western Europe mentioned skills and insight that would increase his country's interoperability with the United States; he stressed the importance of "improved English language skills (written and spoken). An international network. A good knowledge of the US doctrine and therefore an ability to plan and conduct operations with US partners." The international officers mentioned that interactions with their fellow international officers were important as well, helping to build professional networks and better understanding of international relations. Several respondents mentioned improved multinational interoperability and coordination in addition to improved interoperability with the United States. An officer from the Americas remarked, "The opportunity to see and discuss different situations and command styles will help me to be a better leader and to understand and be part of the complex solution of problems that concern many countries." An officer from Africa noted that he gained "broader perspective, critical thinking skills, more creativity in arriving at solutions, greater ease in working in multinational environments."

Many of the officers also expressed the expectation that the professional and personal relationships that they had made with fellow officers would endure throughout their careers and, at the personal level, beyond their careers. Networks and friendships were cited by a senior-level officer from Central Asia who listed "advance knowledge + friends all around the globe + contacts with many industries" as important impacts. The endurance of networks was mentioned by a senior-level officer from Western Europe who remarked, "It has given me an understanding of U.S. culture, the U.S. military decision-making architecture and enabled me to establish some strong friendships that will last long beyond attendance at [the war college]."

Based on the survey evidence, attendance at U.S. military war and staff colleges has a tremendous impact on the international officers' future careers. Because of the high propensity of the officers to be promoted into the top ranks of the militaries, it is also very likely that the officers will have a good deal of influence on decision-making in their future careers. The survey responses also bear out the claim that those military officers who attend U.S. military war and staff colleges form an epistemic community of upwardly mobile military professionals who have personal friendships and professional connections to both their U.S. counterparts and their fellow international officers.

5.6. SUMMARY

U.S. military educational exchange programs are more than a classroom and a curriculum; they are a life experience. In this chapter, survey data from international officers attending U.S. military war and staff colleges were used to learn about the officers' backgrounds and to assess factors that are important in the model of socialization presented earlier. The international officers came from around the world and from different cultural backgrounds. They had previous experience in multicultural and multilingual settings and were not "novices" to international politics. The officers indicated that they were warmly welcomed and extensively integrated with their U.S. military counterparts in the military school and that they and their families were warmly embraced within U.S. communities. The majority of the officers had a significant amount of social interaction with their fellow U.S. officers, not just in the school environment but also in their off-duty time. These interactions helped to counteract negative stereotypes that the officers had of the United States that they had learned by watching U.S. news media and Hollywood movies.

It is clear that the exchange programs educate rising elite leaders from around the world. The chapter discussed four measures that assessed the promotion potential for the officers who attended U.S. military war and staff colleges: first, the level of job they will have immediately after graduating from the U.S. military school; second, the number of people they will have under their command when they return home; third, the type of jobs that prior graduates of U.S. military schools now had; and fourth, the officers' own thoughts on the impact the U.S. school would have on their own careers. The data clearly show that the international officers are very likely to be promoted and occupy important positions in their home countries. Previous graduates were already occupying "very important" positions. Additionally, the data show that within many countries there are substantial cohorts of U.S. military war and staff college graduates who know each other. These upwardly mobile officers have formed an epistemic community within their country that maintains professional and personal contacts with a larger international epistemic community of military officers who understand U.S. military doctrine and share similar experiences as U.S. war and staff college alumni.

Families were a "force multiplier" that enhanced the cultural and social experience for the officers. In particular, the officer's children facilitated a wide range of social and cultural interactions that led to enormously positive (to quote one of the officers) perspectives of their time in the United States. One of the key findings from this chapter is the importance of the entire family's experience to the officer's perception and attitude toward his time abroad. The single most important way that the U.S. government could improve the exchange experience, both in terms of the officer's experience and in terms of accomplishing the program's goals, is to provide financial support so that more officers could bring their families with them.

The data presented in this chapter confirm that the U.S. military war and staff colleges provide an excellent environment for political socialization. But it should also be kept in mind that the officers were selected by their governments to represent their countries and their countries' points of view. Socialization research suggests that the officers' previous experience coupled with a high level of international exposure would tend to make them more impervious to political socialization because they are not "novices" to foreign cultures or international politics. The next chapter explores this issue by assessing the socialization outcomes of the officers' experiences in the United States in the form of the ideas, attitudes, and perceptions of the United States and its citizens that the international officers (and their families) formed while on their exchange program in the United States.

NOTES

1. This survey would not have been possible without the help of many people, and I am grateful for their help, advice, and enthusiasm. I would like to give my immense thanks to the U.S. National Defense University, the U.S. Marine Corps University, the U.S. Army War

College, the U.S. Army Command and General Staff College, and Air University. The U.S. Navy declined to participate.

2. Permission to survey human subjects was granted by Institutional Review Boards at each of the military schools and the author's home institution.

3. There is no universal agreement on what a reasonable response rate should be. Baruch (1999: 434) suggests 36% +/- 13 for senior management personnel and 60% +/- 20 for "rank and file." Babbie (2007: 262) states, "A review of published social research literature suggests that a response rate of 50 percent is considered adequate for analysis and reporting. A response of 60 percent is good; a response of 70 percent is very good."

4. The Institutional Review Boards required that respondents should remain anonymous. If a respondent specified his country of origin, it would be possible to uniquely identify him by name. As a condition for the survey, it was agreed that no comparisons between schools would be made.

5. Munck and Verkuilen (2002: 11) offered this critique. The Polity IV Project is also quite clear that their index excludes civil liberties; see Marshall et al., *Users' Manual* (2011: 14).

6. The Polity IV rating index takes into consideration: (1) the competitiveness of executive recruitment, (2) openness of executive recruitment, (3) constraints on the chief executive, (4) the regulation of participation, and (5) competiveness of political participation.

7. *The Economist* published its first democracy ratings in 2007, and their index is based on five factors: (1) electoral process and pluralism, (2) civil liberties, (3) the functioning of government, (4) political participation, and (5) political culture.

8. In order of increasing repressiveness these countries were: Madagascar, Kuwait, Mauritania, Morocco, Jordan, Ethiopia, Fiji, Burkina Faso, Cuba, Bahrain, Nigeria, Togo, Algeria, Cameroon, Comoros, Niger, Gambia, Belarus, Angola, Kazakhstan, Gabon, Rwanda, Azerbaijan, China, Qatar, Egypt, Côte d'Ivoire, Vietnam, Swaziland, Congo (Brazzaville), Oman, Guinea, Tunisia, Zimbabwe, Yemen, UAE, Tajikistan, Afghanistan, Sudan, Eritrea, Syria, Djibouti, Democratic Republic of Congo, Laos, Guinea-Bissau, Libya, Iran, Equatorial Guinea, Saudi Arabia, Central African Republic, Myanmar, Uzbekistan, Turkmenistan, Chad, and North Korea.

9. Johnston (2001: 494) defined socialization as a process undergone by novices; Thies (2003: 547) discussed state socialization as a process undergone by novices; and Checkel (2001) found "novice officials" to be more influenced by argumentative persuasion than are experienced officials.

10. Officers from authoritarian countries spoke on average 2.9 languages; those from democratic countries spoke on average 2.6 languages.

11. Officers from authoritarian countries traveled to an average of 4.3 countries; those from democratic countries traveled to an average of 5.8 countries.

12. The mean number of countries visited for the respondents who had traveled was 5.8 countries.

13. The breakdown was: 1 from Asia-Pacific, 3 from Africa, 3 from the Middle East, 3 from South and East Asia, 4 from Eastern Europe, 4 from North, South, or Central America.

14. As determined by two-tailed Z-test for difference in proportions at the 0.05 significance level, figure 5.5 shows that this difference is driven mostly by the difference in the proportions of officers who had stated that they spent more than half their time with U.S. people.

15. All of the quotations from the international officers are presented verbatim except for minor spelling corrections as well as material in brackets [ ] that I used to clarify or replace

material that would identify a specific individual, specific school, name of the officer's country, or name of a specific individual.

16. This question was not asked of the international students at the Air Command and Staff College (ACSC); it was asked in the surveys at all of the other schools. The ACSC survey was the only survey that asked the question: "Based on your experiences this year, would you recommend other officers from your country bring their family with them?" 100% of respondents recommended that officers bring their families with them.

17. The region identified as "North, South, or Central America."

18. I thank Becca McBride for this insight.

19. These officers remarked that the impact on their careers would be "decorative/ceremonial," "some but not much," and "no impact." No officer thought it would have a negative impact.

20. At the time of the survey, 5% of respondents had not yet received their new job assignments. Another 2% that I have labeled "Other" are officers who indicated they would be faculty members at a military school or military academy or had accepted a civilian job.

21. Difference in proportions tests found no statistically significant difference between the two groups.

22. Only two of the officers indicated that they did not know someone else from their country who had graduated from one of the U.S. military war or staff colleges. This might be because they did not know of a previous graduate, or they could have been the first officer from their country to attend the school. There were also five respondents who did not answer the question.

6

Perspectives and Opinions of the Exchange Officers

This chapter continues the assessment of the impact of the military educational exchange experiences on its participants. Chapter 5 described the officers' backgrounds, future prospects, and professional and social interactions with fellow U.S. officers and ordinary U.S. citizens. This chapter turns to examine the knowledge, beliefs, attitudes, and perceptions that the exchange participants in the eight different military schools acquired during their time in the United States. What lessons did they take away from their experiences? And, what do these lessons imply for U.S. military soft power?

To summarize the findings in this chapter, the survey data show that much like exchange participants the world over, the international officers at U.S. military war and staff colleges come away from their time abroad with increased understanding and warm feelings toward their host country and the people who welcomed them into their lives and communities. This was true for officers from nondemocratic countries as well as those from democratic countries. Therefore this chapter demonstrates that the military educational exchange programs are a source of soft power for the United States.

While this finding may seem intuitive in retrospect, it is not clear *ex ante* that this should be the case. As was demonstrated earlier, the international officers were certainly not "novices" to foreign cultures, propaganda, or international politics. They were handpicked by their home governments for their loyalty to those governments, and they are likely to hold important positions in those governments when they return home and as they progress in their careers. Unlike young high school or college students who go abroad on exchanges, the international officers have substantial ideological and professional ties to their home governments, they are vested within their home political systems, and they view their American experience as a tool to advance their careers. Thus they should be relatively impervious to socialization attempts. The international officers come to the United States to gain military knowledge and military expertise; the expectation is that they should most appreciate and remark upon the military expertise that they gained in the course of their studies in the United States. However, as we will see in this chapter, this was not the case.

For the international officers it was the cultural experience that was most treasured rather than any particular military expertise or subject matter that was taught in the classroom. Many of the officers talked about the friendships that they had made, the hospitality that they had been shown, and the experience of being embedded in American culture as their best experiences. This is important in at least two respects. First, it shows how international exchange programs that emphasize social interactions can be powerful socializing environments. In the case of military exchanges, those participants who should have been most impervious to its effects nevertheless came away with warm feelings toward the host country and its citizens. Second, it demonstrates from the perspective of those being socialized the importance of social interactions within the process of socialization. It is not only knowledge acquired in the sense that people learn new ideas, but also important is how learners experience these ideas within a wider community, and this leads to the acceptance of these ideas. Social interactions are indeed key to socialization; they are also the foundation upon which long-term friendships and social networks are formed. In the case of the international officers, this epistemic community spans geographic borders, diverse lifestyles, and competing political ideologies.

6.1. FRAMEWORK FOR THE ANALYSIS

The analysis in this chapter examines the socialization process as it occurs in the lives of real people. It brings to life the model of socialization discussed in chapters 2 and 3 by examining (1) how the international officers evaluated their experiences while in the United States, and (2) the opinions and perceptions that the international officers took away with them of the United States and U.S. citizens. In order to do this, the international officer survey contained several open-ended questions. Two questions were used to allow the officer to reflect on his experiences and on the impressions and knowledge that he would take home after graduation. These questions were:

1. What are some of the best experiences you (and your family) have enjoyed this year?
2. What is the most important thing that you learned about the United States in the last year?

Two questions were used to elicit both positive and negative evaluations of their experience and aspects of the United States. These questions were:

1. In your opinion, what is the best thing about the United States?
2. In your opinion, what is one thing that the United States could do to improve?

As mentioned earlier, open-ended questions such as these can be very informative because the respondent is not restricted to any set of choices but can write whatever he wants to write. The respondents were free to list or describe as many or as few items as they wanted. They were also free to write nothing at all. Some of the officers wrote a lot and others wrote more sparsely. The officers' own responses were used to identify different categories of responses for the analysis. These categories emerged from the officers' own remarks. That is to say, there were no prepared categories for coding the survey data, but categories were designed based on the survey responses. So, for example, if no one mentioned a concept/item there is no category for it.

In the next four sections, the results for each of the questions listed above are systematically discussed. Additionally, the results for the respondents from nondemocratic countries are compared to those from respondents from democratic countries. In the final section of the chapter, the overall research findings are summarized. This includes an assessment of the impact of the exchange experience on its participants as well as a discussion of the implications for U.S. military soft power.

6.2. BEST EXPERIENCES DURING THE YEAR

In order to assess the impact of different types of experiences on the ideas that the officers had formed about the United States during their year abroad, the survey asked: "What are some of the best experiences you (and your family) have enjoyed this year?" This survey question sought to identify the types of experiences that the international officers considered their most memorable and, hence, some of the perceptions of the United States and its citizens that the officers would take home as a result of their exchange experiences. Most of the officers identified several best experiences; some respondents described their best experience in detail; others with only a few words.[1] Up to four experiences were coded for each respondent. As mentioned above, the officers' own responses were used to identify different categories of experiences from their narratives. Table 6.1 shows the aggregated results.

The results were categorized based on the activities and memories represented by the responses, reflecting in many cases the wording used by the officers themselves. Yet, these categories were also intertwined. The respondents frequently cited interlinked aspects of their experiences. For example, the military colleges integrate the international officers with their U.S. counterparts in seminar groups. The U.S. officers typically organize study groups as well as social events; friendships between seminar members' families also provided the opportunities for the international officer's families to observe U.S. culture and facilitated travel around the local area. An Eastern European officer who noted many different aspects of his program as his best

Table 6.1. Best Experiences

Best Experiences	*Times Cited*	*% of Items Cited*
Travel in the United States*	120	24
U.S. cultural and social experiences	115	23
Making friends	84	17
War or staff college aspects	58	12
International cultural experiences	38	8
Family aspects	29	6
Social events	25	5
Other	22	4
	491	99%

* This category does not include field trips that were organized by the school. Field trips were cited by 29 respondents and are included under the category "war or staff college aspects."

NOTE: Numerous respondents cited two or more experiences. Percentages do not equal 100% due to rounding

experiences illustrates this meshing of experiences: "Very informative curriculum of [the school]. Nice, healthy, warm relations with US and other countries citizens. Great opportunity to see different parts of US thanks to trips organized by [the school] (both, military and civilian experiences)." Likewise, an officer from Western Europe described his best experiences as: "Two-fold: 1. Being able to have my family with me for the year. . . . 2. Living in a different country long enough to experience the 'true' culture and not just the tourist 'view.'" An officer from Africa also cited many best experiences: "Engaging myself on the issues linked to American culture, gathering with American families, visiting amazing places throughout the U.S., improving English skills." And, the same was true for an officer from the Middle East who listed "the weather, friendly USA people, and Civilization. But the thing is I can say my opinion in anything whatever it is without hesitation."

Personal travel around the United States was clearly the single item that was most frequently mentioned by the officers (line 1 in table 6.1). One hundred twenty survey respondents mentioned traveling or trips that they had made on their own or with their families as one of their best experiences. The survey response of an officer from Africa illustrates this: "We enjoyed travelling, we had the opportunity to see many places that we dreamed to visit, such as Washington, DC–NY–Niagara Falls–Disney." An officer from South and East Asia wrote that his best experiences were "travelling in some states in the U.S. Seeing many unique and interesting/new things which we do not find in our country. Getting along with the Americans, to speak English and to know them better." The respondents mentioned many travel destinations, but the places most mentioned by name were Disney World and Orlando, Florida (mentioned by twenty respondents), New York (mentioned by sixteen respondents), and Washington, DC (mentioned by thirteen respondents). As noted in the last chapter, for the officers with small children, their trips to Disney World were sometimes described as one of the highlights of their year abroad. Other officers mentioned Disney World in conjunction with the family's other memorable experiences; such as the officer from Western Europe who wrote, "My children have wonderful memories from Disneyland, which we visited, but I and my wife more liked American kindness and friendly attitude. We tried to visit as much as possible famous places around Alabama and also inside."

In addition to family and personal trips, twenty-nine respondents cited the field trips that were organized by the schools as part of their official curriculum as one of their most memorable experiences.[2] In the survey responses, the official field trips were mentioned not only for their specific educational focus, but also for how they helped the officer to understand U.S. people, U.S. history, and U.S. culture. For many of the officers, traveling around the United States was a way to get to know the "real" America and experience its society and culture firsthand whether that was on an officially organized trip or on family vacations. Travel also provided a new point of reference for considering (and reconsidering) how the people that they met during their travels compared to previous impressions that they had learned about Americans from American television shows, Hollywood movies, and other sources. Several officers remarked upon the fact that one of the most important things that they learned while in the United States was that their interactions with U.S. people helped to dispel negative stereotypes that they had held prior to their exchange program.[3]

U.S. cultural and social experiences were the second most frequently mentioned "best experiences" (see line 2 in table 6.1), with the closely related categories of making friends (line 3 in table 6.1), international cultural experiences (line 5 in table 6.1), and social events (line 7 in table 6.1) also mentioned by many respondents. These categories included comments on experiences within U.S. culture, the friendships made by the international officer and his family with U.S. people and with other international officers and their families, participation in various social events, interactions with sponsors, and the hospitality shown by U.S. people. As an illustration, an officer from South and East Asia wrote, "I enjoyed travelling in the U.S. especially when participating with U.S. friends. My seminar often had a party on Friday or Saturday evening. We got to know each other so much better compared with a limited amount of time we had in [another school]."

The importance of making friends and memories of specific social events can be seen in the remarks of an officer from Asia-Pacific who wrote that his best experience was how "our family could meet the wonderful sponsors. We spent much time with them; they were like our parents in the USA. We could learn how Americans think, feel, and also we could share common values with them." Getting to know their U.S. counterparts and fellow international officers in the classroom as well as meeting random U.S. citizens was important to how the officers evaluated their year abroad. This can be seen in the answer from a Middle Eastern officer who described his best experience: "To live with an American family for a period of approximately 8 days, and participating with them in most events of the day and getting to know their lifestyle and way of life." An officer from Eastern Europe cited "getting deep insight into American society, sharing thoughts with U.S. and IO students." And, an officer from South and East Asia wrote, "Seeing U.S. society functioning is one of the best experiences for myself and my family. It was nice to know how different U.S. is as is perceived outside U.S."

While interactions with Americans were frequently noted, many respondents also mentioned the friendships and cultural experiences that they had shared with their fellow international officers and their families. For example, as his best experience an officer from Asia-Pacific cited, "Getting to know the other international fellows and forming enduring relationships with a broad number of countries." Taken all together, the remarks of the officers show the importance of social functions and social interactions to the ideas, perceptions, and attitudes toward the United States that travel home with the officer and his family. While the formal classroom instruction is important in facilitating interactions, it is the informal settings and off-duty experiences that form some of the best memories that the officers and their families take back to their home countries.

How do officers who come from nondemocratic countries view their experiences in comparison to those officers from democratic countries? It might be thought that since these officers come from countries perhaps more antagonistic to the "American way of life" they might be more inclined to remark upon the formal aspects of their exchange experience such as learning about U.S. military doctrine and U.S. national security as their best experiences. But this was not the case as can be seen in figure 6.1, which

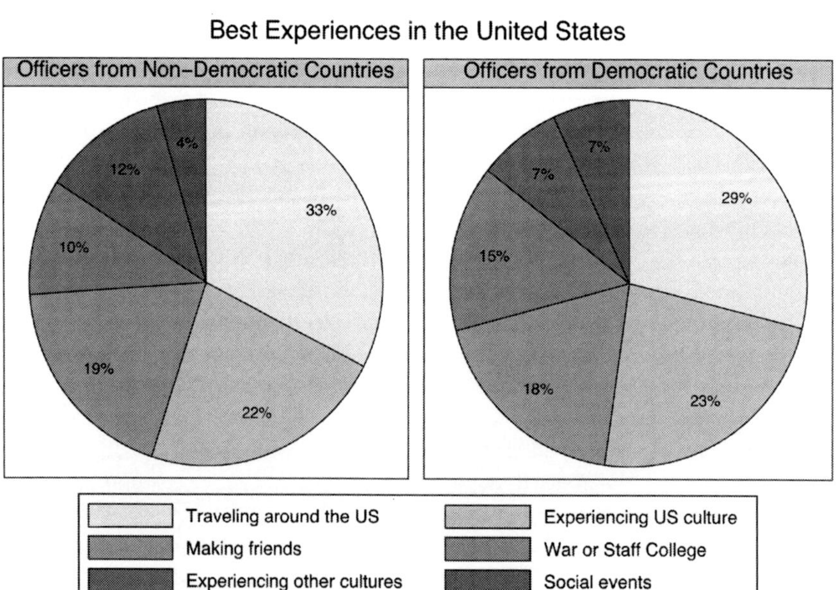

Data from International Officer Survey

Figure 6.1. Best Experiences by Regime Type

shows results broken down by whether the survey respondent came from a nondemocratic country or a democratic country. As can be seen, there is some variation in the types of experiences cited, but none of these differences were statistically significant.[4]

The results in table 6.1 and figure 6.1 show that the officers found memorable experiences in both their formal school curriculum and in life with their families in local communities, yet it was the cultural experiences that come to the forefront as the bases of their best memories. Among the respondents who cited their war or staff college program/curriculum as one of their best experiences (line 4 in table 6.1), twenty-nine respondents cited aspects of their classroom or education program, and (as mentioned above) twenty-nine respondents cited field trips organized by the school.

While the formal experiences in the classroom were less frequently cited as best experiences, it should also be remembered that it is the formal program that helps the officers get to know one another and facilitates other activities that then form the most memorable experiences for the exchange officers. The official school curriculum and activities provided many opportunities for the officers to get to know each other as is evidenced by the thirty-eight respondents who talked about exposure to other cultures and the interactions that they had had with their fellow international officers and their families as one of their best experiences. For example, a senior-level officer from the Middle East remarked that he had a "very good experience at [the military school] especially with the American fellow students and friendship with International Fellow students. My family made some friends in schools." Likewise, an officer from the Americas linked his school experience with his personal experiences: "The best experience was the straight contact with my classmates, on duty and off duty."

It is difficult to determine how much of the cultural experience that the officers remark upon is a direct outcome generated by the programs and curriculum at the military schools and how much is indirect and serendipitous. Certainly the military schools played a significant role because they supported the sponsor programs and provided the classroom environment in which friendships were formed between the officers that were then extended into their off-duty times. A well organized program that helps the officers manage everyday problems, from how to enroll children in schools to how to arrange for utilities at home, is in some sense invisible (or only visible if it functions poorly). It is also a tribute to the U.S. officers who routinely interact with their international counterparts that their international counterparts come away with a very positive experience. The importance of the military community in the officers' experiences was remarked upon by many of the officers, for example, ten of the officers explicitly wrote about their appreciation of the strong sense of community amongst the military families, both U.S. and international officers' families. This sentiment was expressed by an officer from Western Europe who described his best experiences as "the various trips organized by the [military school] to discover the American history and culture. The daily life on base and the sense of community among military families." In a similar response, an officer from Eastern

Europe wrote, "We've had the chance to meet two great families—community sponsor's and barrack sponsor's family, and to celebrate together all the important events of the year." And, another officer from Western Europe remarked, "We have enjoyed being part of the [local] community immensely. Our perception is that we have been looked after extremely well by the local community; there are not many places in the world so welcoming of foreigners."

Inclusion of the officer and his family in social activities and help with the basic tasks of acclimating to life in the United States is important. Under the category "Family aspects" in table 6.1 (line 6 in table 6.1), respondents remarked upon their interactions with other U.S. families, opportunities for children and spouses, interaction with the families of other international officers, and the ability of the entire family to learn/improve their English and experience a foreign culture together. Significant family events that emerged in the survey include at least one international officer who became a new father and at least one who got married during his year in the United States. In considering all these various aspects of the officers' experiences, it is the one-on-one personal interactions that mattered most. While we might think of the military schools as providing regimented military education, the survey points to personal interactions and personal experiences navigating the United States as most memorable. It is not the educational curriculum per se that is important to the process of socialization, but the everyday interactions that occur because the officers have been invited by the U.S. military schools to live for an extended time in the United States. This was true for officers from nondemocratic countries and those from democratic countries. This is a theme that we will follow as we examine further survey responses.

In summary, a response from an officer from the Americas aptly illustrates the overall sentiment expressed in answer to this survey question: "I consider that I and my family had a lot of best experiences. Every week we found something to remember forever." That these are good memories of family opportunities, friendships, travel, and community events is perhaps the most important impact that the military schools have. In the longer term, the attitudes and perceptions gained through these friendships and personal interactions may have a much more profound impact than any formal activity that the international officers participated in. Nevertheless, to reiterate, it is the trials and tribulations of learning, working, and socializing together with their counterparts from around the world within the U.S. military schools that is the environment in which these lasting relationships are formed.

6.3. MOST IMPORTANT THING LEARNED

The last section assessed the types of best memories that the exchange officers take home with them; in this section we turn to look at what the officers described as the most important things that they learned during their year in the United States. In order to understand this aspect of their experience, the survey asked the question:

Table 6.2. Most Important Thing Learned about the United States

	Times Cited	*% of Items Cited*
How Americans think and act	98	29
How U.S. democracy works	59	18
Aspects of U.S. lifestyle and culture	56	17
U.S. military culture, doctrine, and leadership	45	13
U.S. government policies and processes	44	13
U.S. foreign policies and foreign relations	23	7
Other	10	3
	335	100%

NOTE: Numerous respondents cited two or more items.

"What is the most important thing that you learned about the United States in the last year?" This again was an open-ended question where the respondent was free to write whatever he wished. This survey question sought to identify what the international officer thought based on both his formal curriculum within the school and from his everyday life. Whereas the previous question asked for best experiences; this question was neutral in that it asked for important lessons rather than focusing on positive memories. Once again many of the officers identified more than one item.[5] Up to four items were coded for each respondent. Again, the officers' own responses were used to identify different categories of important things learned about the United States from their narratives. These categories emerged from the officers' own remarks. Table 6.2 shows the aggregated results.

Reflecting the neutral nature of the question, the respondents' answers contained both positive and negative evaluations of the United States. Overall, the aggregate tone was positive, and the officers identified quite a variety of things that they learned about the United States and its citizens. Figure 6.2 shows overall results broken down by whether the respondent was from a nondemocratic country or a democratic country. As can be seen, there is some variation in the type of lessons learned that the respondents cited, but none of these differences were statistically significant.[6]

As can be seen in table 6.2 as well as in figure 6.2 insights into understanding how U.S. citizens think and act was overwhelmingly the most important thing that the officers thought they learned about the United States (line 1 in table 6.2). Many answers were fairly general, but some respondents wrote specific comments on the patriotism and commitment to their country of U.S. citizens, the diversity of opinions amongst U.S. people, people's respect for others' opinions, and the hard-work ethic of U.S. people. To illustrate, an officer from Asia-Pacific wrote that the most important thing he learned was "how the U.S. leaders and military officers see the world and other countries' culture. Decision-making process in the U.S." And an officer from South and East Asia wrote the most important thing that he learned about the United States were "core values both for military and civilian. For most civilians that I know, they have the country as their highest

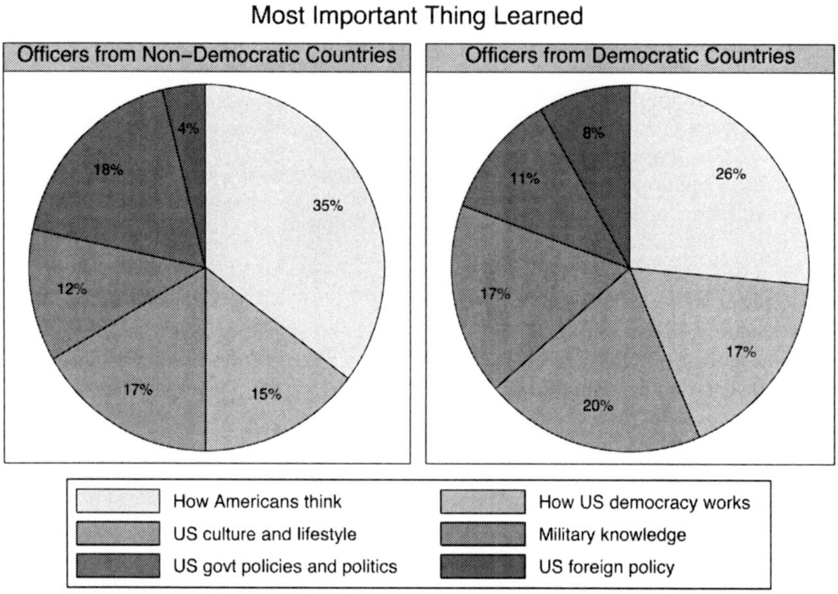

Data from International Officer Survey

Figure 6.2. Most Important Thing Learned during Time in the United States by Regime Type

priority and they sacrifice their lives for the country and their belief which is good based on my perspective. I hope people in my country can be like this." And an officer from the Americas remarked upon "the wide variety of opinions and perspectives held by American officers and citizens, and the profound beliefs of most of its citizens in the fundamental freedoms the U.S. enjoys."

A substantial number of responses contained critiques about how U.S. citizens think and act. For example, fourteen respondents commented on the self-centered and self-interested nature of the United States. Respondents across all regions of the world remarked upon the lack of interest and knowledge of world affairs of U.S. people such as an officer from Western Europe who observed that he learned that in the United States there is "lack of information and correct knowledge of the non-US world." An Eastern European officer wrote, "80% of US citizens do not know much about the world." And an officer from Africa learned that U.S. people "take little interest in international affairs."

U.S. media were seen by some officers as contributing to the ignorance of U.S. people about international events. An officer from Asia-Pacific noted how U.S. thinking shapes and is shaped by U.S. media that "rarely covers anything outside of the United States, unless it specifically relates to U.S. personnel or interests." In some cases the U.S. media were criticized both for their own biases and for leading U.S. citizens astray, as the same officer noted, "Americans tend to have quite a narrow world view + they are heavily influenced by the American media." In a similar observation, an officer from Eastern Europe remarked that "US citizens cannot fully understand the world as it is because they are observing it through their very U.S. lenses." And an officer from the Middle East commented that "people are misled by media."

That the United States was inwardly focused and U.S. people did not pay attention to world affairs is not only a theme that emerged in the answers to this particular survey question, but it is an observation that again emerged in other questions that will be discussed shortly. Several respondents sought to balance their critique of U.S. egocentrism with more positive evaluations of what they learned about the United States, such as an officer from Asia-Pacific who remarked that "the U.S. is still too self absorbed. Despite all the good the U.S. does for the world, the world balance of power is shifting in ways hard to predict. The U.S. must adapt." An officer from the Americas stated, "I learned that I did not really understand the United States. It is a very complex society. I also learned that Americans really do not understand the political, social, and cultural issues at play in other countries."

While the officers were critical of U.S. media, they were also somewhat critical of media portrayals in general. Eleven of the respondents remarked upon their own media-generated biases about the United States, particularly how Hollywood films and news reporting had shaped their image of the United States and U.S. people. These respondents wrote that the U.S. citizens that they had met were not the same as the stereotypes that they had seen in Hollywood movies or other media. An officer from Western Europe wrote that he had learned that "not all the information that I had on the U.S. was accurate and real. The U.S. is slightly different from the perception provided by movies and television." An officer from Eastern Europe observed that the United States "is completely different country than you can assume from Hollywood movies." An officer from the Americas wrote, "The USA is very different than as shown on the Hollywood and television programs and movies." An officer from Eastern Europe wrote, "I have learnt that most of the stereotypes about the U.S. are not true. They are not a pragmatic cold nation as presented in the international media. I very much admire their strive for democracy and better way of life."

The second most frequently mentioned lesson learned by respondents was how democracy works in the United States (line 2 in table 6.2). These answers included observations on the freedoms afforded to individuals, on the nature of civil-military interactions, on how people have equal rights, and that there is accountability of citizens and officials. A key component in establishing democracy is first being able to provide security for citizens. Rule of law to include accountability of citizens and officials is a foundation of U.S. democracy and U.S. society; and it is probably taken for granted by many U.S. citizens. Thus, it might be surprising that quite a few of the officers did mention it. Nine respondents explicitly wrote that they had learned that for the most part U.S. people had respect for the law. Respondents talked about the value gained when people respect the law, are held accountable, and how the rule of law operates in the United States. A senior-level officer from Eastern Europe observed that "rule of law and law enforcement are a great part of American society." A senior-level officer from the Americas wrote that the most important thing he learned about the United States was "the respect and responsibility of the citizen, and the importance and the respect for the military." Another officer from the Americas noted that "the people respect the rules." An officer from Asia-Pacific learned that the "U.S. runs by system, people respect the system." And an officer from the Middle East echoed the above sentiments. He wrote that the most important thing learned about the United States was that it was a "systematic life where almost everybody follows the rules."

The third aspect most commonly mentioned pertained to the lifestyle and culture of the United States (line 3 in table 6.2). Many respondents simply listed U.S. culture as the most important thing that they had learned while other respondents explicitly identified such aspects as the open nature of U.S. society, the existence and acceptance of diversity, and how hard work brings success; as well as more critical observations of how money negatively influences politics and the media. For example, an African officer wrote, "It is a country full of opportunities in terms of increasing your education." An officer from Asia-Pacific remarked that since "all men are created equal so, have a try to unleash your dream." An Eastern European officer learned that "the United States is country which provides for its citizens the most favorable conditions of life." Another officer from Eastern Europe learned that "building standards and priorities are stairs that will climb you up."

But not all was positive. Critical remarks included observations on the importance of money to U.S. citizens and how wealthier people have greater access to education and health care, and how bipartisan politics were dividing and weakening the United States. For example, one respondent learned that "democracy is not easy and there are often times when it feels like United States is tearing apart due to highly divisive views on issues such as health care, abortion, gun rights, etc." Another observed that "if something cannot be logically explained or justified it is called politics."

Many of the respondents discussed specific knowledge that they had gained through their course of study at the military college as the most important thing that they had learned (line 4 in table 6.2). These items included how the Joint Operations Planning Process works, how the U.S. Army is organized and operates, how military operational planning is done, how joint operations are conducted, the nature of U.S. military culture, and lessons about leadership. For example, an officer from Asia-Pacific wrote that the most important thing he had learned was "how to become a better strategic thinker." An officer from South and East Asia wrote that he learned "how to improve every aspect by getting feedback and continuing analyses. How to build a system, law enforcement and self-discipline." A few respondents also singled out aspects of U.S. government policies and processes as the most important thing they learned about the United States (line 5 in table 6.2). These responses were closely related to subjects that the officers studied in the classroom including topics such as U.S. national security policies, U.S. decision-making processes, and U.S. domestic policies. Thus, for a respondent from the Middle East, "the role of the lobbies in shaping the foreign policy of the nation" was his most relevant lesson learned about U.S. policy-making. A respondent from Eastern Europe was more sanguine: "Good state, but not perfect. There is no country in the World that accepts diversity and that people adapt to quickly as it is here. Efforts to provide security for its citizens and the security of the country on the one hand, and the emphasis on the internal disagreements about certain issues that are important to citizens of USA are the dominant themes of today."

Finally, aspects of U.S. foreign policies and foreign relations (line 6 in table 6.2) were cited by twenty-three respondents as the most important thing that they learned. This included general comments such as that made by one officer who wrote that the most important thing that he learned was "what makes U.S. foreign policy tick," as well as observations about U.S. policy toward other regions of the world, the extent to which the United States could or should be perceived as a reliable ally, and the extent to which the United States worked with other countries. For example, an officer from Asia thought that "the U.S. pursues too aggressively its own interest." While an officer from Western Europe found "that the U.S. Government sees that it needs partners and that unilateral actions not necessarily are of the best," while another officer from Western Europe found "enduring congruence of values and interests with my country."

To summarize the findings from this survey question, when they were asked to identify the most important thing that they had learned about the United States, the international officers identified various types of knowledge they gained by living within local communities and interacting with a broad array of U.S. people as well as knowledge that they had gained in their classroom studies. By their own assessment, the most important types of knowledge that the international officers gained were about the cultural and social aspects of their experience and in particular having a greater understanding of how Americans think and act. Once again this is not to say that the subjects taught in the schools' curricula were disregarded or thought unimportant, it is only to say that at the end of their stay, the lessons that the officers thought were the most important were the ones that had to do with their experiences living within America and interacting with their counterparts rather than knowledge they had gained through reading books or attending lectures.

6.4. BEST ASPECTS OF THE UNITED STATES

The last two sections looked at how the international officers rated their own learning experiences while in the United States as well as some of the best memories they hold of their year attending a U.S. military war or staff college. This section examines how the officers evaluated the United States per se. After a year of observing and interacting with U.S. people and U.S. institutions, what do the officers identify as the best and worst aspects of the United States? In order to elicit positive evaluations the following survey question was used: "In your opinion, what is the best thing about the United States?" Once again this question was open-ended, and respondents were

Table 6.3. What Is Best about the United States

	Times Cited	% of Items Cited
U.S. people and culture	84	22
Freedom	65	17
Governance and rule of law	57	15
Democracy	53	14
Economy and infrastructure	44	12
Opportunities	37	10
Diversity	21	6
Educational levels and system	8	2
Other	9	2
	378	100%

NOTE: Numerous respondents cited two or more items.

free to list or describe as many or as few items as they wanted. Up to four items for each respondent were coded. Again, the officers' own responses were used to formulate the categories.[7] Table 6.3 shows the aggregated results.

When asked to identify what they thought was best about the United States, respondents wrote about the country, its ideals, and the people themselves. Multiple aspects of the United States were listed by many of the respondents, such as an officer from Africa who wrote that the best thing about the United States was "being able to express your views, excellent institutional framework, effectiveness of security apparatus and communications systems." An officer from South and East Asia remarked that "the institutions are extremely strong and the respect for law is visible. Equal opportunities for all—discrimination is negligible. The culture of accepting and listening to opposing viewpoints. Taking care of the old, sick and veterans." A respondent from Asia-Pacific wrote, "The interesting thing about America is its progress in various areas. Ability to manage and regulate national life in the democratic system. Natural beauty makes America such an unforgettable hometown."

As can be seen in table 6.3, the aspect of the United States most commonly cited was the U.S. people themselves (line 1 in table 6.3). Under this category, the officers specified characteristics that they appreciated in the U.S. people: these included remarks on friendliness, helpfulness, kindness, hospitality, ambition, optimism, and creativity. These remarks are consistent with comments that the officers had made on other survey questions. In many ways, this result is a testament to the success of programs at the military schools, such as the introductory courses for the officers when they initially arrive in the United States that help them to integrate into local communities and solve problems that they might encounter in day-to-day life. It is also a testament to the local military and civilian communities that welcome the officers and to the many sponsor families that "adopt" the officers and their families. A comment by an African officer illustrates the importance of these programs; he noted that the best thing about the United States was "making all-comers at home and the Americans [officers'] willingness to go out of their ways to help non-American officers." The sense of welcome and inclusion were mentioned by many of the officers and are represented here by an officer who wrote that what he thought best about the United States was "how much the military is appreciated in the public, and how the [international] families are integrated," and another officer who commented that what was best about America was "American patriotism and the respect Americans have for their soldiers."

In identifying the best aspects of the United States, respondents frequently mentioned democracy and freedom—the quintessential ideals that Americans themselves use to characterize their own country. Many of the respondents saw the United States as a free and democratic nation that provides good governance, everyday security, and opportunities for its citizens to improve their lives. Freedom and democracy were clearly linked with rule of law: people could live as they chose because they were secure. One officer commented that in the United States "individual rights are very important and policemen have great responsibility and enormous authority." Another officer from Eastern Europe wrote that what is best about the United States is that "the U.S. gives many opportunities and freedom of personal expression."

Freedom was the second most frequently cited best aspect of the United States (line 2 in table 6.3) and was very closely associated with the respondents' comments on democracy and rule of law. In many cases these three aspects of the United States went hand-in-hand. For example, an officer from Western Europe listed "democracy, democratic accountability, rule of law, ideals of freedom and liberty" as the things he found best about America. The category "freedom" encompassed remarks by the officers on freedom of speech, freedom of the press, and freedom of religion as well as thirty respondents who listed "freedom" specifically. Of the types of freedom specifically mentioned, "freedom of speech" was most frequent with thirteen respondents citing it.

Governance and rule of law (line 3 in table 6.3) were specifically cited by a substantial number of the officers. This category included specific comments on government structure, social services, government organization, and government effectiveness in providing public services as well as comments on rule of law, respect for the law, accountability, and responsibility. For example, an officer from the Americas wrote that what was best about the United States was "The respect for the laws of the country." Within this category (Governance and rule of law), almost half (twenty-seven of fifty-seven) of the comments mentioned rule of law, respect for the law, or accountability as one of the best aspects of the United States reflecting that in the United States people are generally secure in their everyday activities, laws and police are seen as legitimate, for the most part all people are held accountable to the law, and that corruption is less than in other parts of the world. As an example, the many economic opportunities that U.S. people had were linked to good governance by a respondent from Eastern Europe who wrote that for him the best thing about the United States was: "1) The way of the business is developed. 2) The judicial system. 3) Opportunity for personal prosperity." Another officer from Eastern Europe remarked upon "The openness of all governmental institutions and candor when discussing any governmental issue" as what had impressed him during his time in the United States.

Democracy was also specifically cited by a substantial number of respondents (line 4 in table 6.3). This category encompassed comments that the officers made on respect for individual rights, equal opportunity, and that through hard work people can achieve their goals because they had equal rights. Thirty-three respondents listed "democracy" specifically as the best thing about the United States. In respect to the American ideal as a democratic and free nation, some of the respondents remarked upon the fact that it was an ideal that other peoples around the world hope for but have less success in achieving. For example, one respondent wrote that the best aspects of the United States were: "The strength of the institutions—government, military, education, civil services, infrastructure, law enforcement etc. These are very robust and provide a great deal for the citizens, this is something that most countries are unable to provide for their people." An officer from Africa observed of the United States: "The people in this country have the opportunity to manage their lives better than in other countries." Reflecting that the American way of life is an ideal that might also be under pressure in the United States, one officer from the Americas wrote: "It is a modern democracy; their constitution embodies a vision that has led to its greatness as a nation. Depending where you live, you can have a good and prosperous life. I am not sure how much longer that will last."

Respondents also mentioned the opportunities (line 6 in table 6.3) that the United States affords its citizens such as an excellent educational system (line 8 in table 6.3) as well as U.S. economic strength (line 5 in table 6.3) as the best aspects of the United States. For example, an officer from the Americas wrote that the best thing about the United States was "opportunities for the people to express and do what they want looking forward to making a better country and place to live." Ten respondents specifically noted that in the United States success could be achieved by anyone through hard work. An officer from the Middle East wrote that the United States was a "land of opportunity, if you have an open mind you will succeed." This also harkens back to comments the officers had made on accountability, responsibility, and the rule of law that were seen as leveling the playing field for all people regardless of race or political connections. An officer from Africa noted that the best thing about the United States was that it was "technologically advanced and have great people in the right jobs. Jobs are given out on what you know not who you are." The United States was seen by many as the land of opportunities unlike some of the respondents' own home countries. An officer from Asia-Pacific wrote that the best aspect of the United States was "the availability of resources and immense resilience against any crisis." An officer from Eastern Europe noted that in the United States "Everyone has a better chance than in the rest of the world." An officer from Africa wrote that the best thing about the United States was "the fact that everyone has a chance" and another wrote that "everyone does his best in his job."

Respondents also remarked upon the diversity of America as one of its best aspects (line 7 in table 6.2). Under this category, most respondents (ten of twenty-one) simply listed "diversity" without further elaboration. For the other respondents their comments were about three aspects of diversity in the United States: the diversity of its landscape (four respondents), its ethnic diversity (four respondents), and multiculturalism (three respondents). For example, an officer from Eastern Europe remarked that what he appreciated most about the United States was that "it is so culturally and climatically diversified that you can do a lot of things in any given season or period of time." One officer from Western Europe commented on "the vast, diverse, generally clean and fantastic nature," while another officer from the same region wrote that the "friendly nature of the people; huge diversity; the scale of the place" was what he found most impressive.

Figure 6.3 compares the perspectives of military officers from nondemocratic countries to those of officers from democratic countries.

The perceptions and attitudes of officers from nondemocratic countries differed from their counterparts from democratic countries in two significant ways. First, officers from democratic countries were more likely to comment on U.S. people and U.S. culture as the best aspects of the United States than were officers from nondemocratic countries. Second, officers from nondemocratic countries were more likely to appreciate and remark upon aspects of rule of law in the United States.[8] Only a few of the officers from nondemocratic countries elaborated their answer in any detail, but rather most of them listed items such as "rules and law," "rule of law," "accountability," and "respect for law." A few comments were particularly revealing. One officer from a nondemocratic country, for example, wrote that the best thing about the United States was that it was "a society that holds everyone accountable, responsible but at the

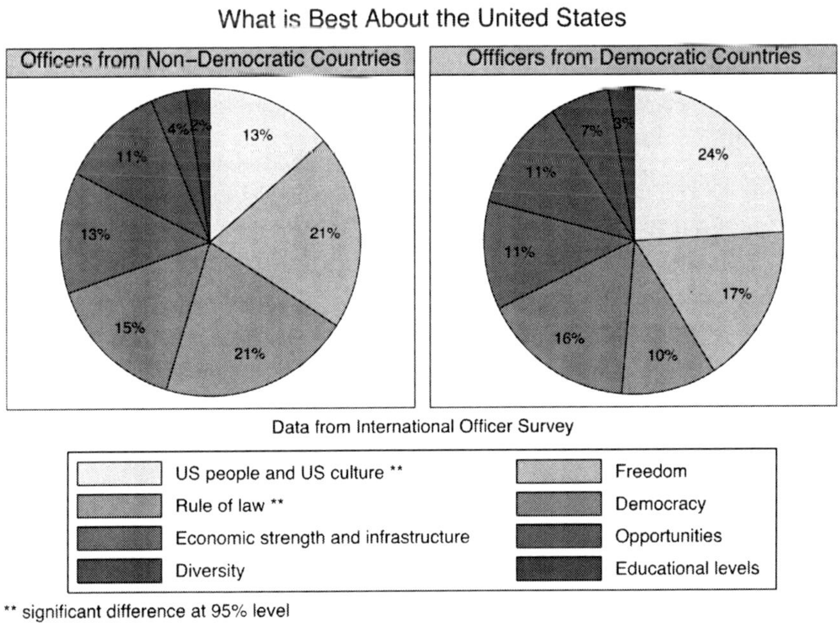

Figure 6.3. Best in the United States by Regime Type

same time everyone has rights and privileges that he enjoys." Another officer from a nondemocratic country wrote that the United States "is a good example of democracy and a best example where different people of color can share the same interest in a country without major conflict or massacre." These remarks would indicate that rule of law was seen as a necessary part of democratic rule that allowed people from diverse backgrounds equal access and equal opportunities regardless of political connections. From this perspective it seems logical that this aspect of the United States would be appreciated more by officers from countries where autocratic rule was rather arbitrary and autocratic leaders were supported through family connections, corruption, and the lack of accountability of those in authority positions.

To summarize, the United States, its citizens and its government, self-referentially portray themselves as the "land of opportunity," the "home of the free and the brave," and the land "with liberty and justice for all." As seen in table 6.3 and in the above discussion, this was a characterization that could be drawn from the international officers' experiences and perceptions as well. Indeed, one officer wrote that in evaluating the United States "probably the best thing is the well known cliché, 'the land of opportunity.'"

Despite the U.S. unilateral invasion of Iraq, the ongoing war in Afghanistan, the extrajudicial holding of prisoners at Guantanamo Bay detention facility, and the George W. Bush Administration's approval and use of torture (Bush 2010: 170–71), this idealistic image of the United States continued to resonate with the survey respondents, many of whom came from countries not allied to the United States and where anti-American sentiments are claimed to be prevalent amongst the general population. This positive perception of the United States, however, was also balanced by the belief of a great many of the international officers that the United States and its citizens were self-centered and did not know and did not care about the world beyond their borders. This is a topic that we turn to in the next section.

6.5. HOW THE UNITED STATES COULD IMPROVE

In order to understand the negative perceptions of, attitudes toward, and beliefs about the United States and its citizens the international officers might take away with them, the international officers were also asked: "In your opinion, what is one thing that the United States could do to improve?" The aggregated results are shown in table 6.4.

The international officers generally took away with them a positive impression of the U.S. people with whom they had interacted. However, while the officers expressed good opinions of these interactions per se, they were nevertheless critical of U.S. people's knowledge of the rest of the world; several found the United States in general to be culturally insensitive and arrogant in the world arena. An extended comment by one officer illustrates this point in a poignant way. He wrote that the United States should work to build "a greater awareness of the broader international community. The U.S. is very American-centric and many of the U.S. citizens I have

Table 6.4. What the United States Should Work to Improve

	Times Cited	*% of Items Cited*
Respect for other cultures and countries	71	27
Knowledge of the rest of the world	67	26
U.S. foreign relations and diplomacy	43	16
U.S. societal problems	21	8
U.S. health, education, and social services	19	7
U.S. domestic policies and politics	18	7
U.S. media bias	12	5
Environmental awareness	7	3
Other	4	2
	262	101%

NOTE: Several respondents cited two or more items. Percentage does not equal 100% due to rounding.

interacted with believe that most of the world owes them for their security and economic status. Along with these issues is a low appreciation of the richness of other cultures. The U.S. is a wonderful place but there are also many other wonderful places outside the U.S."

In table 6.4 the respondents' answers were categorized based on ideas expressed by them. The labels for the first three lines were based on phrases frequently used in their answers. The first focuses on respect for others, the second on knowledge about others, and the third on the conduct of the United States abroad. The comments encompassed within these three categories focused on a broader theme, specifically the respondents' perception that the United States and its citizens were egocentric, focused on themselves, and had little knowledge or concern for peoples outside of their own borders. As can be seen, approximately 70% of the responses were of this nature. This perception was held by respondents from around the world and not focused in any one region. In terms of respect for other cultures (line 1 in table 6.4), twenty-seven respondents specifically said that the United States needed to show greater respect for other countries and other cultures, and another sixteen respondents thought that the United States needed to show greater empathy for other peoples and listen to what they had to say. For example, an officer from the Americas urged that the United States should "understand in a real way what are the conditions of the other people, how different it is to try to accomplish the work without resources. Hear, and stand in the shoes of others."

In terms of knowledge of U.S. people about the rest of the world (line 2 in table 6.4), thirty-one respondents specifically mentioned that the United States and its citizens should work to improve their knowledge of the rest of the world and another twenty-seven respondents remarked that U.S. people needed to learn about other countries and cultures. An officer from Asia captured a common sentiment among many officers when he urged that the United States should work to "cultivate a sense of caring for others even though different nations and languages." The lack of empathy toward other cultures and peoples is also a sentiment that has been remarked upon by U.S. citizens and U.S. military officers, so these types of comments certainly have validation within a segment of U.S. society. As one example, in a recent award-winning essay on lessons learned while deployed to Afghanistan, a U.S. military officer wrote, "I learned that language is not only a barrier between us and the host nation, but also between us and our coalition partners. A great asset is lost because of our lack of patience with our non-English speaking partners" (Stattel 2010).

Across world regions the international officers thought that both the United States and the world as a whole would be better off if U.S. people were better informed about the rest of the world (line 2 in table 6.4). An officer from the Middle East wrote that the United States should: "1—understand more others perspectives. 2—be more open to friends and allies experiences and feedback specifically on cultural, political and related military issues concerning their countries or countries of interest for the good of all." These sentiments were echoed in the comments of an officer from Western Europe who also noted the consequences of not working to build understanding and respect for others. He wrote that the United States should work more to "build effective, trustful, genuine partnerships. You have a lot of partisans: don't make them turn their back to you. It is not in our respective interest." Another officer from Western Europe added historical perspective: "Try to forget about your role of Cold War lonely super-power and work as a Great Power amongst allies and other Great Powers."

Harkening back to previous comments on media bias in other parts of the survey, twelve respondents thought that U.S media did not have good coverage of international news and was exceedingly biased, which contributed to the ignorance of U.S. people about the rest of the world (line 7 in table 6.4). Respondents remarked on the need to reform the U.S. media that they thought misled and misinformed those who watched it. One officer called for the "need to instill greater sense of social responsibility within the media to report in an unbiased nature." He went on to say that U.S. people and U.S. society needed "to engage with the media to ensure the content does not incite fear, hate or create degradation of people's values."

The highly technological nature of U.S. military capabilities was seen by some of the officers as a hindrance to developing more substantial personal working relationships with U.S. partners. One respondent wrote that the "U.S. military is technologically very dependent organization. Nevertheless, working with people from other cultures sometimes requires sincereness and trust. It is very difficult to achieve that with technological approach through mechanistic manner, check-lists, percentages and measures of effectiveness."[9] Another respondent wrote that the United States needed to improve its "work with others (countries) respecting their pace/needs."

The survey responses indicated a clear desire that the United States should work together with others in the international community and be better informed about international events. For example, one officer from Asia wrote that the United States needed to improve "the way it interacts with others" and "should learn that the 'U.S.' way of life may not be the best or most suited or liked by the world." Another officer from Africa thought that the United States needed to improve "multilateralism" and "have a deeper understanding of the culture of others and avoid general interpretation." Other comments urged the United States to show greater humility, respect the sovereignty of others, be more diplomatic, not stereotype Muslims, and show greater awareness for global environmental problems.

It is important to emphasize that the comments concerning the stereotyping of Muslims came from several regions and not solely or even predominantly from the Middle East. This indicates a broader perception amongst the international officers that the United States is at worst anti-Islamic and at best ignorant of Islam. Specifically in respect to U.S. attitudes toward Islam, an officer who was *not* from the Middle East wrote that the United States should not "generalize that all Muslims are the same" and should understand that many in the Muslim world "share the same value of living together and in harmony as the people of the U.S." Another officer who was *not* from the Middle East wrote for the need to have "objectivity in the media and avoid a stereotype vision of the Muslims (Islamophobia)." In respect to cultural tolerance more broadly, one officer from the Americas wrote that the United States should work to "develop a greater understanding of the differences of those from other countries and a willingness to accept that while their beliefs may be different they are no less valid then those cherished by the typical American." In relation to the idea that the United States is an exceptional nation, an officer from Africa quite elegantly observed that "American exceptionalism is OK. Americans would do much better in international engagements if they had greater sensitivity to the fact that other cultures have pride too. Exceptionalism need not result in contempt or disrespect of other cultures. The U.S. could educate its population better about other cultures, or at least about the fact that other cultures do have a basis for the values they uphold, even when they differ from American values."

Interestingly, some of these same sorts of criticisms that the officers had made in the international context about the United States were also offered about U.S. domestic politics, which were seen as divisive rather than constructive (line 6 in table 6.4). For example, an officer from Asia-Pacific observed, "The extent of polarization in politics can damage the activities of the government." For some respondents, U.S. citizens were seen as lacking interest, caring, and knowledge about their own political system. And an officer remarked that U.S. citizens needed to choose "the right people in the government" rather than "only people looking for their selfish benefit." One of the respondents observed that U.S. people needed to get to know each other better. He noted the need for "improvement of sensitivity and the relationship of people towards one another, first in the neighborhood and then spreading. People between themselves are increasingly moving away and thus they do not know each other well despite the fact they live in the same neighborhood for years." Another respondent thought that the United States was on an inevitable decline: "The current political, economic and social situation is now leading [the United States] down a path of decline. Within one generation much of its national power will have been squandered. It will be a divided society. I do not see much improvement in this prognosis."

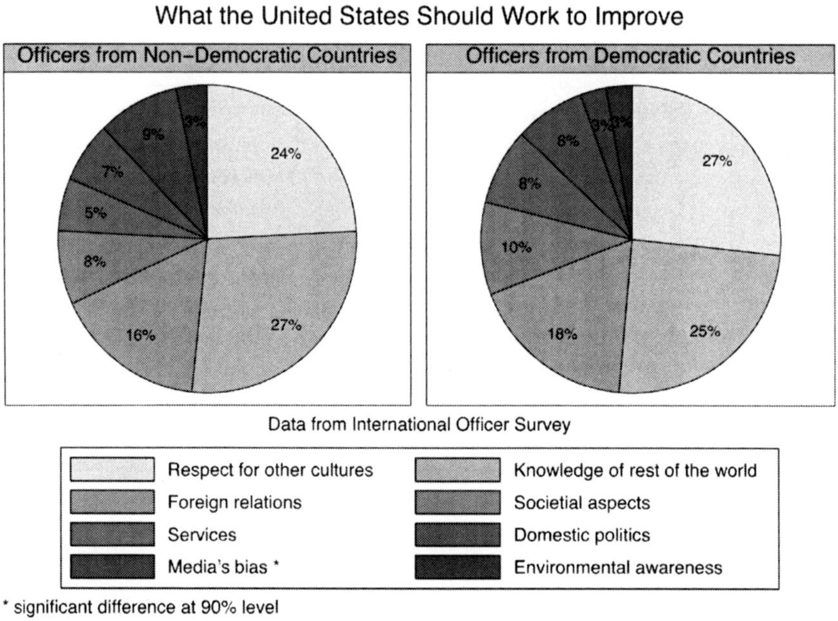

Figure 6.4. What the United States Should Work to Improve by Regime Type

Figure 6.4 compares the perspective of military officers from nondemocratic countries to those from democratic countries.

The perceptions and attitudes of officers from nondemocratic countries did not differ much from their counterparts from democratic countries with one exception. A significantly greater proportion of officers from nondemocratic countries were more likely to say that the United States needed to work to eliminate bias in U.S. media. That being said, the comments between the two groups were quite similar. Most of the respondents only listed "media" as something that the United States needed to improve. Those officers who elaborated thought that U.S. media should provide better international coverage. This, of course, reiterates the widespread belief amongst many of the officers that U.S. people should know more about the rest of the world.

6.6. SUMMARY

The survey results in this chapter showed how soft power is built through the socialization of exchange participants from the perspective of those involved and what the outcomes of a socialization effort look like in real life. At the micro-level of socialization, people acquire new ideas through classroom learning, observation, persuasion, and from social interactions. Many of the respondents remarked directly on how their own ideas of the United States had changed; for example, quite a few officers wrote that they had come to the United States with preconceived notions and stereotypes of the United States and U.S. people, but that the experience of living and traveling in America had changed those perceptions for the better. Reflecting on the impact of the exchange program on his military career, an officer from Western Europe remarked, "I don't know about my career but my understanding and approach is changed significantly."

Once again, it should be emphasized that it was not just the officers as individuals who were influenced by their year in the United States, but also the officers as the heads of their own families. The family, and in particular the officers' children of, were key to building positive memories; as an officer from the Americas observed: living in the United States created "more awareness about the U.S. and the American people. The experience definitely changed [my family's] view in a positive way, about the American people and institutions." Similarly, an officer from Eastern Europe wrote that his family had "already changed the assumption about US in a positive side. The American experience will help them feel more comfortable and independent in our society. They will want to come back to the U.S. again." These observations illustrate what soft power looks like when expressed by individual people.

In the lives of people, experiences and outcomes are rarely of the "ideal type" around which theories are formed. One of the themes that emerged from the responses to the survey questions is the sophistication of the officers' views about the United States. This is evident in their appreciation of the complexities and contradictions in U.S. society and politics. For example, an officer from Asia-Pacific wrote that the most important thing he learned about the United States was that it was "a land of contrast and contradiction." Another officer from Western Europe wrote the most important thing he learned about the United States was "that the typical American does not exist; the array of differences is even larger than I expected." In this, we can see Seymour Martin Lipset's (1996) idea of American Exceptionalism come to life in how the officers think about the United States and in the ideas that they have formed during their year in America. As an African officer remarked, to learn about "American exceptionalism" was important to understanding the United States. Because the United States was seen as a nation like no other, as an exceptional nation, it was held to high standards and sometimes disappointed those who also found value in its people and government. This observation was not just inferred from the various responses of the officers, but was also directly stated by them, as the extended remarks by one officer on American Exceptionalism discussed earlier showed.

The officers found much to admire but also much to criticize in the United States. Sometimes the same facets of American society are both esteemed and criticized; sometimes by the same respondent. The officers' multifaceted views of the United States are consistent with recent scholarship on world public opinion of the United States in more general settings (Chiozza 2009). The responses illustrate the nature of the perceptions and memories that the officers take home with them. Overall, the international officers found American society to be safe, orderly, and full of opportunities. People were free to express their views and live their lives as they chose. Although better opportunities were available for rich Americans, nevertheless because people had equal rights, they could work hard to achieve their goals and dreams. This was an opportunity that many wished more people outside of the United States could have. Rule of law was seen as a necessary part of democratic rule that allowed people from diverse backgrounds equal access and equal opportunities regardless of political connections. From this perspective it seems logical that this aspect of the United States would be appreciated more by officers from countries where autocratic rule was rather arbitrary and autocratic leaders were supported through family connections, corruption, and lack of accountability of those in authority positions. In this sense, the officers from nondemocratic countries saw that another way was possible and their responses indicated that they aspired to this ideal.

The international officers were critical of the self-centeredness of the United States and the lack of knowledge of so many Americans about the world beyond the borders of the United States. This ignorance was attributed to lack of interest and biased media reporting. The United States was seen as powerful and its citizens had opportunities and freedoms, but at the same time, the United States and its people were also seen as self-centered and ignorant of life in the rest of the world. Because American society was generally seen as egalitarian and American justice was seen as less arbitrary than in other parts of the world, the international officers regretted the lack of awareness of U.S. citizens about the rest of the world because, they felt, the whole world would be better off if U.S. people were better informed and held their politicians accountable for formulating foreign policies that could help others who were less fortunate.

The findings in this chapter demonstrate how the U.S. military, its officers, and schools are important agents through which the United States might develop its own image and, hence, its own soft power throughout the world by engaging their counterparts worldwide. While formal education at the military school consumed much of the officers' time, it was the personal interactions on a day-to-day basis that shaped the positive attitudes toward their U.S. counterparts, both military and civilian. These memories went home to many countries around the world. Overwhelmingly, these were positive memories of the friendships the officers had formed and of how warmly they had been received by both officials at their schools and the ordinary Americans that they met. These were friendships and social connections that the officers hoped would last for many years to come.

I am particularly grateful to the international officers who participated in the survey. Their insights and comments are a valuable contribution to our understanding of socialization processes as they occur in the lives of real individuals. Their insights also help us to understand how the military schools shaped their perspectives of and attitudes toward the United States. In considering the categories of best experiences that emerged from the responses of the survey participants it seems clear that many of the memories that the officers took home with them were of the social activities and social interactions that they had. The fact that many of these social interactions are remarked upon as "best experiences" illustrates their importance in shaping the officers' positive evaluation of their year in the United States. The comments also show how friendships were formed that linked the officers with their U.S. counterparts, other U.S. citizens, and their fellow international officers. These links exist on multiple levels: between the officers, between spouses of the officers, and between the children of the officers.

The responses also showed how relationships were built between U.S. military officers and their foreign military counterparts from around the world—much of it facilitated through travel, family involvement, official and unofficial social events—all while living life in U.S. communities. While the international officers did express clear criticisms of the United States and its citizens, these comments existed in conjunction with a deep appreciation of the opportunity to attend a prestigious U.S. military school and gratitude to those people who helped them during their stay in the United States. The final item on the survey stated, "Please write any other comments you might have about your experience this past year." This was an open-ended statement with no indication that any response was expected. Yet over half (137 of 244) of the officers took the time to write something. Most of the comments were very similar to those expressed by the following four officers. From a respondent from Eastern Europe: "One of the best years of my life." From a respondent from Africa: "It was an awesome year that I will never forget." From a Middle Eastern respondent: "Was a great opportunity to interact with US military officers of different diversity. I learnt a lot from them as they learned from me. I have met and established great friends. The friendships will last for a long time." From an Asian respondent: "I have learned a great deal from this experience. It has been very enjoyable for me and my family. Thank you so much."

NOTES

1. One hundred forty-one respondents (58%) identified more than one item. Thirteen officers did not answer the question; they were fairly evenly spread across regions and most (75%) were from democratic countries.
2. Personal trips were cited by one hundred officers, unspecified travel was cited by twenty officers, and trips organized by the school were cited by twenty-nine officers. Field trips are included under the category "war and staff college aspects."
3. I discuss this in more detail in the next section of this chapter.
4. Difference in proportions tests showed no statistically significant differences in response rates for all categories at the 90% or greater confidence levels.
5. Seventy-seven respondents (32%) identified more than one item. Twenty-one officers did not answer the question; they were fairly evenly spread across regions with sixteen coming from democratic countries and four from nondemocratic countries.
6. Difference in proportions tests showed no statistically significant differences in response rates for all categories at the 90% or greater confidence levels.

7. Ninety-six respondents (39%) identified more than one item. Seventeen officers did not answer the question; they were fairly evenly spread across regions with thirteen coming from democratic countries and two from nondemocratic countries (two of the respondents who did not answer this question also did not rate the level of democracy in their country).

8. Difference in proportions tests showed a statistically significant difference between the proportions of respondents citing U.S. people and U.S. culture and rule of law. There were no statistically significant differences in the proportions of respondents for all other categories.

9. Interestingly, one U.S. veteran of the war in Afghanistan wrote, "Though the best of intentions may be from where we start, I have come to realize our objectives are almost always lost in metrics and want for numbers" (Stattel 2010).

III

IMPACT OF MILITARY EDUCATIONAL EXCHANGES ON DEMOCRATIC TRENDS

7

Impact on the Development of Democratic Institutions

Since World War II, the United States has sought to place its military at the center of a network of militaries from countries around the world. The goal of this effort has been to build military partnerships that will enhance U.S. security and create a stable and secure environment where the United States and other countries can pursue their interests without resorting to war. An important part of this effort by the United States has been to support and promote the adoption of democratic institutions and practices by countries the world over. U.S. national security strategies across political parties and generations have all explicitly stated that promoting and supporting democracy is a core value and objective of the United States.[1] Military-to-military exchanges are an important component of this strategy. But, to date, there has been very little empirical data on the long-term impact of the exchanges on participants' countries. This chapter and the next provide empirical evidence that the military exchanges not only serve as a source of information and soft power, but they also support U.S. foreign policy efforts to build and sustain democratic governance around the world.

The analyses in this chapter use historical data covering the period 1972–2006 to examine the extent to which the educational exchanges hosted by the United States' war and staff colleges impact long-term trends in democratization. The chapter begins with a discussion of the importance of promoting democratic governance in U.S. foreign policy, and, hence, the importance of finding effective strategies. The chapter's analyses use hazard models to assess the extent to which the educational exchange programs hosted by the U.S. military's war and staff colleges have helped or hindered the development of democratic governance. The modeling results show that countries whose political leaders sent their military officers to study at U.S. military war and staff colleges were more likely to develop and consolidate democratic political institutions than were countries that did not send their officers. These findings hold while controlling for a large array of pretreatment confounders, meaning potential factors that influence participation in the exchange programs as well as the outcomes on the dependent variable. The democratizing result occurs whether or not it was the intention of the political leaders; in fact, the effect is strongest in those countries whose leaders have no desire to democratize, the "hardcore" authoritarian countries.

7.1. DEMOCRATIZATION AS A SECURITY STRATEGY OF THE UNITED STATES

U.S. policymakers as well as U.S. citizens have traditionally viewed the export of democratic ideals, values, and institutions as beneficial to the peoples who are "democratized" as well as an outcome that in the longer term contributes to a more peaceful international system and promotes U.S. national security. Normative and pragmatic beliefs in the efficacy of U.S. democratization efforts have deep ideological and historical roots. These roots can be traced back to ideas put forth by Immanuel Kant in *Perpetual Peace: A Philosophical Sketch*. This treatise provided the theoretical foundation that linked liberal beliefs in individual freedom and democratic governance with realist concerns for national security and national interest. Published in 1795 when the United States was still in its infancy, Kant argued that international peace might be obtained through a federation of republican states. The treatise serves today as the conceptual foundation for arguments supporting both scholarly and policy beliefs that democratic states form the basis of a more peaceful international system because democratic states rarely, if ever, use large-scale combat operations against each other to resolve political disputes.

Scholars Melvin Small and David Singer (1976), R. J. Rummel (1983), and Michael Doyle (1983a and 1983b) were largely responsible for the reemergence of Kant's argument in contemporary form. Their arguments spurred an extensive research program on the role of democratic governance in preventing war. Despite criticisms of the democratic peace theory, the finding that democracies do indeed resolve their conflicts with each other short of war has held up to rather intense scrutiny and serves as the foundation for contemporary U.S. foreign policies that seek to promote democracy abroad.

In addition to this theoretical argument, the preference of the United States for an international system of democratic states was also shaped by historical lessons learned, particularly from the great world wars of the early twentieth century. Four days before the United States declared war on Germany in 1917, U.S. President Woodrow Wilson justified U.S. involvement by linking peace with the furtherance of democratic governance:

> It is a distressing and oppressive duty, Gentlemen of the Congress, which I have performed in thus addressing you. There are, it may be, many months of fiery trial and sacrifice ahead of us. It is a fearful thing to lead this great peaceful people into war, into

> the most terrible and disastrous of all wars, civilization itself seeming to be in the balance. But the right is more precious than peace, and *we shall fight for the things which we have always carried nearest our hearts—for democracy, for the right of those who submit to authority to have a voice in their own Governments, for the rights and liberties of small nations, for a universal dominion of right by such a concert of free peoples as shall bring peace and safety to all nations and make the world itself at last free.* (Wilson 1917, italics added)

Yet, in the aftermath of the war, the United States failed to pursue these same goals effectively. Collective security proved ephemeral when the United States declined to join the League of Nations. Without active U.S. involvement, ideals of self-determination evolved into the rise of nationalistic authoritarianism in Europe during the 1930s. It was only after the second cataclysmic world war rooted in such authoritarianism had destroyed much of Europe and a large portion of Asia that U.S. efforts to promote democratic governance became firmly established and acted upon in U.S. foreign policy. In the aftermath of World War II, the United States recognized the need to rebuild a Europe that was democratic, free, and economically open; as well as the need to counter and contain the expansion of the Soviet Union. Democratic governance, individual freedom, and free market economies were seen as the liberal counter to the autocratic rule that was being imposed on many countries by the United States' great nemesis, the Union of Soviet Socialist Republics.

National security scholars Amos Jordan, William Taylor, and Michael Mazarr (1999: 55) have argued that U.S. foreign policy has always been shaped by the messianic sort of idealism that is reflected in Wilson's speech to Congress, but that this idealistic view of the American way of life became even more deeply embedded in the American psyche in the aftermath of World War II: "America had become a status quo power, its people essentially satisfied with life as they knew it, holding their condition of peace and harmony at home as an ideal for all rational people everywhere. They believed in the virtues of democracy and took it for granted that the fruits of democracy should represent meaningful goals to all people throughout the world." Indeed, within U.S. society and U.S. foreign policy discussions, democracy was asserted to be good not only for American citizens, but for other peoples and for the international system as a whole. For example, U.S. President George H. W. Bush exhorted his national security establishment: "Let us work together to lead the world toward the 21st Century, the Age of Democratic Peace. There is no more important goal to which we could aspire" (U.S. White House 1993: 21).

While U.S. leaders and citizens had a rather idealistic view of American ideas of democracy, such ideals also had a sound realist basis. As noted above, scholars provided empirical evidence that security was enhanced for all states within the community of democratic states because war was nearly unheard of between democratic countries. Taking Kant's treatise and scholars' findings to heart, U.S. leaders reasoned that as countries adopted democratic institutions, they were much less likely to use military force to resolve disputes with the United States, a fellow democracy. So in this respect, enlarging the community of democratic states promotes U.S. national security interests. This is not to say that U.S. foreign policymakers have single-mindedly pursued democratization. Particularly in the Cold War era, U.S. affinity for democracy was balanced by important national security concerns and did not prevent the United States from supporting autocratic regimes as a counter to the expansion of the Soviet Union. The intermingling of an ideological belief in the efficacy of democratic governance with realist national security concerns continues to be a fundamental aspect of U.S. security strategies that not only seek to "promote democracy abroad" but also, as Ikenberry and Kupchan (1990: 289) would say, to legitimate American hegemony.

This longstanding normative and pragmatic preference for an international system of fellow democratic states has been clearly and explicitly stated across recent U.S. administrations. The George W. Bush Administration, despite its propensity for unilateral actions, argued in its first national security strategy that democratization was a worthy goal to be pursued. The 2002 *National Security Strategy of the United States of America* published by the Bush White House bluntly states, "We will extend the peace by encouraging free and open societies on every continent" (U.S. White House 2002: opening statement, no page number). The tone and content were similar in the second Bush Administration, which argued, "The survival of liberty at home increasingly depends on the success of liberty abroad" (U.S. White House 2006: 3). While playing down the word "democracy" and elaborating more on its rationale, the 2010 national security strategy statement of the Obama Administration expressed much the same sentiment:

> The United States supports the expansion of democracy and human rights abroad because governments that respect these values are more just, peaceful, and legitimate. We also do so because their success abroad fosters an environment that supports America's national interests. Political systems that protect universal rights are ultimately more stable, successful, and secure. As our history shows, the United States can more effectively forge consensus to tackle shared challenges when working with governments that reflect the will and respect the rights of their people, rather than just the narrow interests of those in power. (U.S. White House 2010: 37)

In the United States, scholars, policymakers, and citizens may agree that democratization is a worthy goal to pursue, but the practical problem is that traditional uses of military power to impose democracy are no longer politically feasible, politically effective, or normatively appropriate as a means to achieve this goal. The U.S. military has certainly been used to impose regime change in such diverse authoritarian states as Germany, Japan, South Korea, Panama, and more recently in Afghanistan and Iraq. As these cases show, however, such a strategy can be very costly in terms of money and lives, and in some cases such as Afghanistan, not effective. A democratization strategy based on the socialization of political elites is more efficacious and more cost effective than using military force in its traditional roles. To paraphrase both Sun Tzu and Joseph Nye: to win one hundred victories in one hundred battles is not the acme of foreign policy skill; it demonstrates your lack of power over events because you were forced to go to this extreme measure; to convince others to want what you want is the best strategy because it extends your influence, costs less, and is the height of foreign policy skill.

7.2. MILITARY EXCHANGE PROGRAMS AND DEMOCRATIZATION

The analyses in chapter 6 showed the effectiveness of military exchanges in building goodwill toward the United States. The exchange officers also developed an appreciation for why U.S. people valued democracy and individual freedoms. In many ways the officers' remarks on the various aspects of democracy in the United States should come as no surprise because one of the U.S. government's goals in hosting the officers at U.S. military war and staff colleges is to "increase the ability of foreign military and civilian personnel to instill and maintain democratic values and protect internationally recognized human rights in their own government and military" (DoD/DoS *Foreign Military Training* 2011: II-2).

The officers who participate in the U.S.-hosted military exchanges come from countries across the political spectrum, from democracies and autocracies, and from those countries where governance has been less stable over time. The governments who send their military officers to the United States to study also have a variety of motivations and goals. For leaders in established democracies, the exchanges are a way to strengthen their alliance with the United States and for their officers to acquire technical knowledge relevant to their careers. For political leaders in struggling and nascent democracies, the exchange programs offer a way for these leaders to give their military officers the opportunity to acquire skills, knowledge, and (importantly) the motivation to help their countries build stable democratic institutions. However, also participating in the exchanges are military officers from what we might call the "hardcore" authoritarian countries whose leaders want no part of the "democratization" enterprise. In these cases, there is no rational reason why civilian or military dictators would send their military officers to an environment to purposefully inculcate new ideas that are antithetical to their maintenance of power. In these cases, dictators send their military officers to the United States to acquire technical skills, useful contacts, and military expertise. However, an unintended consequence of this "technical" education is that the officers experience a new social and professional environment. Any socialization effects that do occur come about despite the best efforts of authoritarian leaders to resist and counter them by choosing only officers who are loyal to the dictator and thought impervious to U.S. influence.

Yale Richmond's (2003) research on international exchanges during the Cold War provides insight into the role played by educational exchange programs in a gradual and largely unobservable democratization process that occurred in the Soviet Union despite extensive actions by Soviet leaders to prevent it. Thus, his study also provides insight into how U.S. military exchanges might impact the authoritarian countries that participate in them. Similar to the military officers who come to the United

States from authoritarian countries, Richmond (2003) showed how liberalization gradually occurred despite the Soviet Union's efforts to prevent it. Richmond (2003) found that educational exchanges increased Western influence, particularly within the Russian intelligentsia, which helped to foster gradual liberalization in the Soviet Union, contributing to the eventual collapse of the Soviet system.

In considering the impacts of the military educational exchange programs, there is no reason to assume that a democratic transition will be overt and dramatic; indeed, as Richmond showed, transitions themselves may be the result of an accumulation of many small acts (and non-acts) of many people that occur over a long time. Chances are that military officers cannot or do not wish to act on any new ideas in any sort of spectacular fashion such as open dissent. However, their experiences while in the United States and the friendships that they formed remain with them in the longer term. Also over the longer term, more officers return home as graduates of U.S. military schools, forming a cohort of officers who have seen how many aspects of democracy work from civilian control of the military to rule of law. In a gradual process these officers may have a long-term influence on the political institutions of their home countries. This may be in prominent actions by the military such as refusal to support repressive actions of their governments or open support for democratic opposition, but more likely it is in a less spectacular fashion such as not taking an active role in civilian political struggles or hindering democratic transitions. From their experiences in the United States, the military officers have also seen that military organizations maintain influence and respect within democratic governments, and this helps to lessen their fears of losing institutional power and personal esteem.

To emphasize, the effect of the military exchanges may not be readily observable as a specific event. For example, the officers' motivation might result in a "non-event" such as in Tunisia in 2010–2011 where the military did not interfere with protesters, but it might also result in an overt, observable event such as in Syria where the military has been used to violently suppress protesters. In evaluating the effects of the military programs, the tendency in the past has been to focus on the readily observable event that receives a lot of media attention, such as in the case of Syria, and to discount the unobservable event, such as in the case of Tunisia, where the military supported democratic changes *by their inaction*. And, it should not go without noting—the Tunisian military has been an active participant in U.S.-hosted exchanges, but the Syrian military has not. It is the accumulated effects of both types of events, overt and unobservable, that the analyses in this chapter and the one that follows evaluate.

7.3. DEFINING AND MEASURING DEMOCRATIC INSTITUTIONS

In order to assess a country's level of democratic governance, it is first necessary to define what is meant by democracy and to have some measure of the various stages of democratization. In this chapter democracy is conceptualized as the presence of political institutions, such as elections, that allow citizens to participate freely in their own governance.[2] In this conceptualization a democratic country is one that has institutions and established procedures through which citizens can choose their political leaders and people can express their preferences for different policies. There are also institutions that oversee and regulate how political leaders exercise their power; the elected leaders cannot do whatever they want but must remain responsive to their constituents because of these institutions. Conversely, in autocratic countries, leaders are appointed from within an elite ruling group or they simply seize power. An autocratic country has few institutional restraints on how leaders exercise their power, and autocratic leaders can pursue their own agendas with little regard for the opinions, welfare, or desires of their citizens.

In order to measure this concept of democracy, the polity rating data from the Polity IV Project was used to define three types of countries: coherent democratic countries, coherent autocratic countries, and the "middle ground" countries that have characteristics of both democracy and autocracy. According to this definition a coherent democracy is one where "(a) political participation is unrestricted, open, and fully competitive; (b) executive recruitment is elective, and (c) constraints on the chief executive are substantial" (Marshall et al. 2011 *User's Manual*: 15). A coherent autocratic country is one that severely restricts political participation of its citizens. Its leaders are chosen from within the political elite by other political elites, and there are very few restrictions or institutional constraints on how leaders exercise power.

Conceptually, coherent democracies are at one end of a spectrum of political identity, and coherent autocracies are at the other end. At the democratic end of this spectrum, Great Britain, Canada, Norway, and New Zealand are examples of "hardcore" democracies; and at the autocratic end, Saudi Arabia, North Korea, the former East Germany, and Swaziland are examples of "hardcore" autocracies. Along the spectrum between these two types of political identities are the countries that have some characteristics of democracies and some characteristics of autocracies. In general, "middle ground" countries are less stable in terms of maintaining either autocratic or democratic institutions; their governance has fluctuated over the years in terms of the extent to which they are able to foster and maintain democratic political institutions.

7.4. MODELING DEMOCRATIC TRANSITIONS

Cox proportional hazard models were used to assess the influence of U.S. military educational exchanges and other relevant factors on the ability of states to acquire and maintain democratic political institutions. The analysis encompassed 169 countries, virtually all countries of the world with a population of at least a half million. The models showed how different factors (called covariates) were associated with a country's history of governance, defined in terms of political institutions as described above.

Cox models are used in a variety of scientific disciplines to assess factors leading some entities to survive longer than others, or conversely, to assess factors leading to premature death or failure. In medicine, hazard analysis has been used to evaluate how a patient's inherent characteristics, such as gender, weight, or genetic predisposition, when combined with external therapies, such as the use of a new drug, affect whether the person lives or dies. In engineering, hazard analysis has been used to understand how environmental stresses, construction techniques, and use patterns lead to equipment failure. For the analysis of democratic governance in this chapter, Cox models were used in a similar manner to assess the extent to which the military exchanges (along with other relevant factors) affected the ability of different types of governance (i.e., democracy and autocracy) to endure over the long term.

In order to model the impact of the military exchanges on the durability of democracy and autocracy over the longer term, a master dataset of 5,148 observations in the form of country-years was divided into three political identities: coherent democracies, coherent autocracies, and middle ground; using the polity definition shown in table 7.1.

Table 7.1. Political Identity Groups

Type of Political Identity	*Polity Definition*	*Number of Country-Years*	*Percent of Total*
Coherent Democracy	polity score > 16	1,789	35%
Coherent Autocracy	polity score < 4	1,806	35%
Middle Ground	4<= polity score <=16	1,553	30%
Total		5,148	100%

The observations were fairly evenly distributed between the three political identity groups. Within the data there were countries that entered the study after 1972 such as Ukraine and Kazakhstan, which were formerly republics of the Soviet Union. There were also right-censored observations; for example, a country that ceased to exist during the time of the study such as Czechoslovakia (which voted itself out of existence in 1993) or countries in existence when the study terminated in 2006.[3] While the total number of countries varied from year to year, most of them (74%) existed for the duration of the period 1972–2006.[4]

Cox models were estimated to evaluate four types of transitions between the three different political identities. In the analyses, there are three types of political identities (coherent democracies, coherent autocracies, and middle ground) and four ways that a transition might occur. The first type of transition occurs when a country "fails" out of the coherent democracy group because its polity score drops below 17. This is an autocratic transition because the country is no longer defined as a coherent democracy, but one with some features of autocracy. The second type of transition occurs when a country "fails" out of the middle group because its polity score drops below 4. This is also an autocratic transition because the country has taken on more autocratic institutions and now belongs within the coherent autocracy group. The third type of transition occurs when a country "fails" out of the coherent autocracy group because its polity score rises above 3. This is a democratic/liberal transition because the country is no longer defined as a coherent autocracy, but one with some features of democracy. The fourth type of transition occurs when a country "fails" out of the middle group by obtaining a polity score of 17 or greater. This is also a democratic transition because the country has further institutionalized its democratic practices and now belongs within the coherent democracy group.

Modeling the impact of the U.S. military exchanges on democratization relies on an extensive and well-developed research program in which scholars have done considerable work identifying the key factors associated with democratization. Factors that scholars have determined to have important effects on a country's propensity to democratize are: its level of economic development, its openness to trade, its religious affiliations, its level of ethnic diversity, whether it was formerly a colony, whether it is a new country, its history of past authoritarian transitions, the Soviet Union's influence on it, the amount of U.S. economic assistance provided, and whether the U.S. military has intervened in it. These are the factors that should capture most of the variation; that is to say, they should determine the extent to which a country is democratizing or becoming more autocratic. Thus, the military exchange variable assesses whether participation in the military exchanges makes a difference while controlling for the baseline trends captured by these other factors.[5] In this, the models do exactly what noted democratization scholar Adam Przeworski has called for as a needed improvement on past studies:

> We can match on observables. But should we not worry about unobservables? Suppose that leaders of some countries go to study in Cambridge, where they absorb the ideals of democracy and learn how to promote growth. Leaders of other countries, however, go to the School for [sic] the Americas, where they learn how to repress and nothing about economics. Dictatorships will then generate lower growth because of the quality of the leadership (. . .) Since this is a variable we could not observe systematically, we cannot match on it. And it may matter. Conditional mean independence—the assumption that unobserved factors do not matter—is very strong, and likely to be often false in cross-national research. (Przeworski 2007: 161)

7.5. RESULTS

Figure 7.1 shows the bivariate effect of U.S. military exchanges on the longevity/durability of each of the political identities. These Kaplan Meier graphs provide a visual representation of how the probability of survival of each political identity is affected by participation or nonparticipation in U.S.-hosted military exchanges over time.

The top two graphs show that autocratic and middle ground countries that participated in the military exchanges had less chance of survival than their counterparts that did not participate. That is to say, that those autocratic and middle ground countries that participated in the military exchanges were more likely to acquire democratic

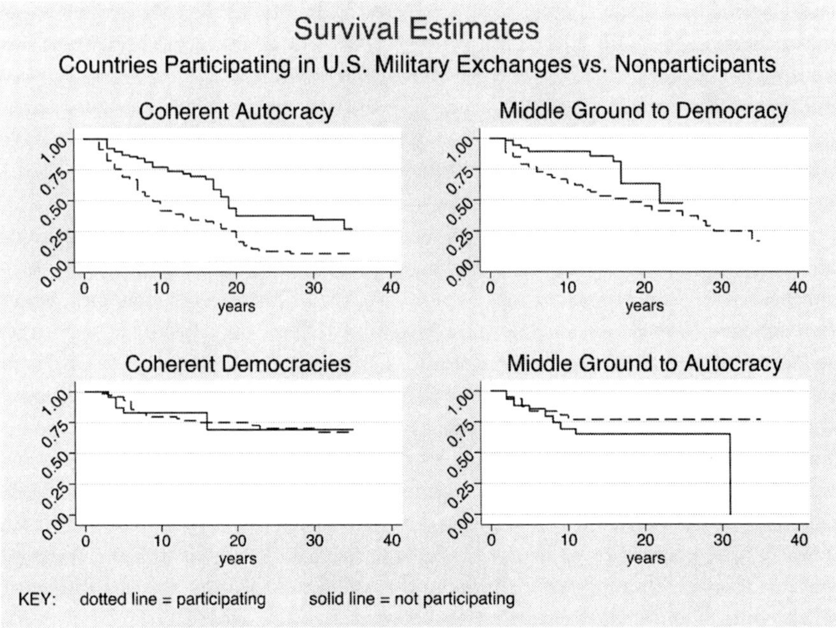

Figure 7.1. Effects of U.S. Military Exchanges on the Longevity and Durability of Different Political Identities

political institutions than were their counterparts that did not participate in the exchanges. In the lower two graphs it can be seen that, consistent with the multivariate analyses discussed below, participation in U.S.-hosted military educational exchanges had no significant impact on democratic countries; the military exchanges neither helped nor hindered these countries maintain their democratic institutions.

The multivariate empirical analysis consisted of four Cox models, one for each transition process discussed above. Both graphical and statistical diagnostics were used to test and validate model specification.[6] Figures 7.2 and 7.3 show the Cox model results, with coefficient values for each model covariate with their accompanying 95% confidence intervals. Figure 7.2 contains the two sets of analyses in which "failure" was defined as a democratic transition.

In these two models a negative coefficient indicates that the associated factor reduced the hazard of transition to democracy and a positive coefficient indicates a factor that increased the hazard of transition to democracy. Thus, a negative coefficient indicates a factor that inhibited democratic transition whereas a positive coefficient indicates a factor that favored democratic transitions. So, for example, in the coherent autocracy group, countries that were relatively new were more likely to remain coherent autocracies than were their counterparts that were not. As can be seen, in both sets of analyses (for coherent autocracy and middle to democracy), those countries that had participated in U.S.-hosted military educa-

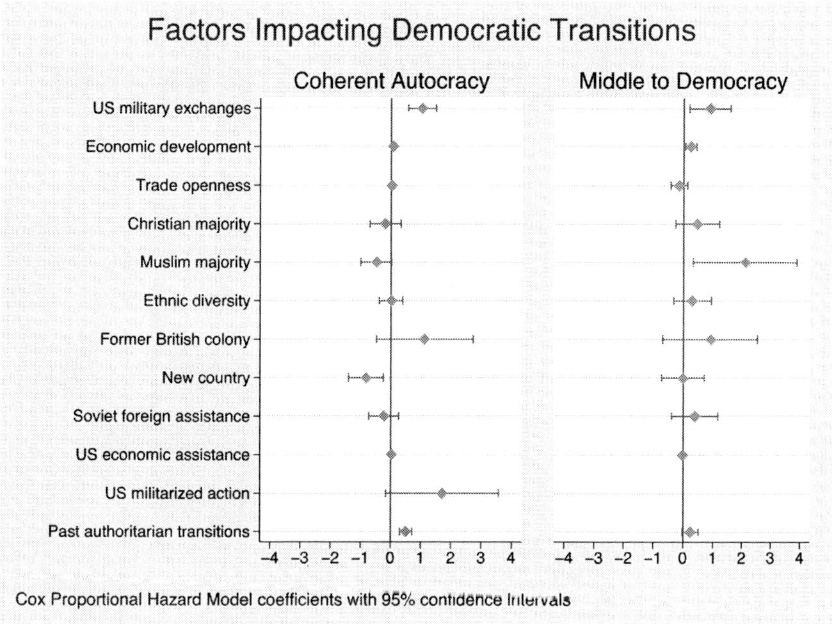

Figure 7.2. Cox Model Results, Factors Impacting Democratic Transitions

tional exchanges were more likely to transition to more liberal political institutions than were their counterparts who did not participate.

Figure 7.3 contains the two sets of analyses in which failure was defined as an autocratic transition.

A negative coefficient indicated that the factor reduced the hazard of failure to autocracy and a positive coefficient indicated a factor that had increased the hazard of failure to autocracy. Thus, a negative coefficient indicates a factor that aided democratization whereas a positive coefficient indicates a factor that inhibited democratization. So, for example, in the coherent democracy group, countries that had a high level of economic development were more likely to remain coherent democracies than were their counterparts that did not; that is to say, this factor had a positive impact on democratic trends. In the middle to autocracy group, the positive and significant coefficient for Soviet foreign assistance means that this factor increased the hazard of failure to autocracy, and thus had a negative impact on democratic trends. In both of these two groups (coherent democracy and middle to autocracy), U.S. military educational exchanges had no significant impact.

Table 7.2 contains a summary of results of the four Cox proportional hazard models in terms of the effect of each of the factors on democratic trends. Consistent with democratic peace theory, the United States did not use military force against other democracies and the coefficient for U.S. militarized actions could not

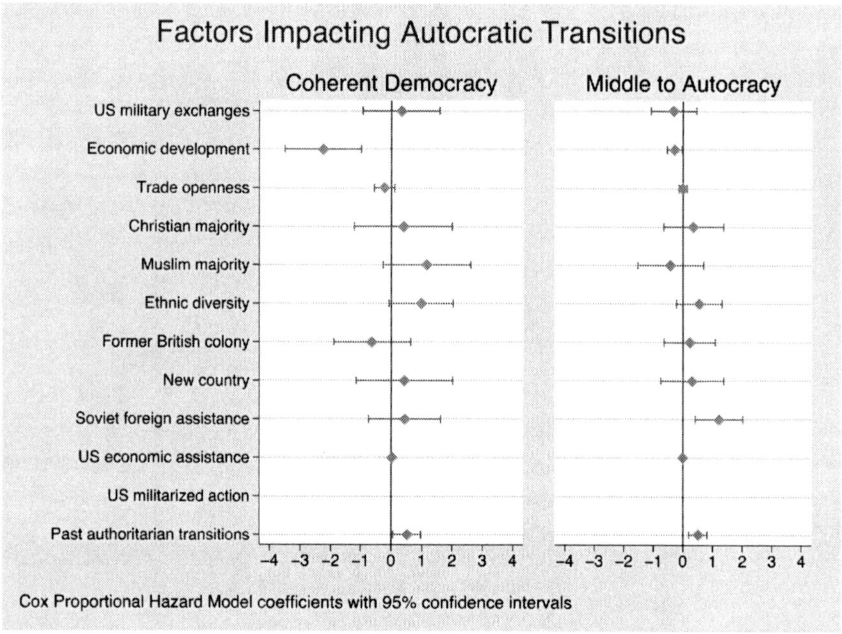

Figure 7.3. Cox Model Results, Factors Impacting Autocratic Transitions

Table 7.2. Summary of Effects of Military Exchanges on Democratic Trends

	Coherent Autocracy	*Middle Ground*	*Coherent Democracy*
U.S. military educational exchanges	Transition +	Transition +	0
Level of economic development	Transition +	Transition +	Consolidate
Trade openness	0	0	0
Christian majority population	0	0	0
Muslim majority population	0	Transition +	0
Ethnic diversity	0	0	0
Former British colony	0	0	0
New country	Consolidate	0	0
Soviet foreign assistance	0	Transition –	0
U.S. economic aid	Transition +	0	0
Previous authoritarian transitions	Transition +	Transition –	Transition –
U.S. militarized actions	0	NA	NA

.05 significance level
"Consolidate" means increases the longevity of this political identity.
"Transition +" means aids in transition to more liberal/democratic identity.
"Transition -" means aids in transition to more autocratic identity.
NA indicates not applicable to this model; not used.

be estimated for middle ground and coherent democracies and the covariate was dropped from these models.

The results show that U.S. military educational exchanges often help democratization, and at the worst, do not hinder it. This result holds while controlling for important factors that scholars have identified as important influences on democratization. As past research would suggest, a country's level of economic development is a key factor influencing its ability to build, support, and maintain democratic political institutions. The impact of economic well-being is felt across the spectrum of countries from the most hardcore of autocratic states to the most hardcore of democratic states. A country's past history was also important. Those countries that had experienced several authoritarian transitions (such as coups) were less stable and more likely to revert to autocracy than those countries that had no past history with such events. As can be seen, U.S. military educational exchanges were positively associated with democratic trends, with the most pronounced effect for coherent autocratic countries. These effects were substantial. Coherent autocratic countries that sent their military officers to U.S. military war and staff colleges were 2.8 times more likely to undergo a democratic transition than autocratic countries that did not send their officers.[7] And, in a cascading fashion, middle ground countries that sent their officers were 2.4 times more likely to transition to coherent democracies than their counterparts that did not send their officers.[8] South Korea, Mali, Paraguay, and Portugal are a few examples of countries that participated in U.S. military exchanges and experienced this cascading effect of democratization.

It is also important to note that the military exchanges had no significant effects in either aiding or hampering the consolidation of a coherent democratic identity. In many ways this finding is not surprising and is consistent with the arguments put forth in this book. Democratic ideas and values are already deeply embedded within the "hardcore" coherent democracies; these are the countries that have already internalized and institutionalized democratic norms and practices.

7.6. SUMMARY

The analyses in this chapter highlight the importance of U.S.-hosted military educational exchanges as an effective strategy to support and promote democratic governance around the world over the longer term. In analyses that spanned 169 countries over the years 1972–2006, the military exchanges were found to be significantly and positively associated with the growth and flourishing of democracy in the countries that participated relative to those that did not participate. The analyses showed that those autocratic countries whose political leaders sent their military officers to study at U.S. military war and staff colleges were more than twice as likely to develop and consolidate democratic political institutions than were their counterparts in other countries who did not send their officers. This effect occurred whether or not it was the intention of the political leaders; in fact, the effect is strongest in those countries whose leaders have no desire to democratize, the "hardcore" authoritarian countries.

NOTES

1. The national security strategies of Presidents Reagan, G. H. W. Bush, Clinton, G. W. Bush, and Obama all explicitly state that promoting democracy is a goal of the United States.

2. In the chapter that follows, democracy is conceptualized in terms of the presence of rule of law and the upholding of basic rights and freedoms of citizens.

3. Cox models are excellent for analysis of data containing right-censored observations; for discussion of this issue see Box-Steffensmeier and Jones (2004: 16–19).

4. A complete list of all 169 countries and the years they were under observation is available from the author.

5. The rationale for the inclusion of each factor in the models is based on a very large body of scholarly work on democratization. This information, as well as information on data sources and covariate coding, is available from the author.

6. These were tests of the proportional hazard assumption as recommended by Hosmer and Lemeshow (1999) and Therneau and Grambsch (2000). Scaled Schoenfeld residuals plots were constructed and examined for all covariates when the covariate's individual test for the proportional hazards assumption indicated that the covariate might exhibit some time dependency. Visual tests were important in judging to what extent the functional form of the individual covariate was, if at all, time dependent. A few covariates did exhibit significant time dependency, and the models were reestimated including an interaction term of the covariate with time, as recommended by Box-Steffensmeier and Zorn (2001: 978). The global tests of the proportional hazards assumption were well within tolerance.

7. The hazard ratio for the covariate *U.S. military educational exchange* was 2.818 in the coherent autocracy model.

8. The hazard ratio for the covariate *U.S. military educational exchange* was 2.442 in the middle ground to coherent democracy model.

8

Impact on Democratic Practices

The analyses in the previous chapter confirmed the positive, long-term impact of U.S.-hosted military exchanges on one aspect of democracy: the ability to build, support, and consolidate democratic political institutions. This chapter focuses on another aspect of democracy: whether political leaders respect the basic rights and freedoms of their citizens. For this analysis multilevel longitudinal models with random effects were used to evaluate the extent to which participation in educational exchanges at U.S. war and staff colleges influenced the extent to which a country's political leaders respected their citizens' ability to exercise their basic rights and freedoms over the longer term. The analyses show that countries that sent their military officers to study at military institutes in the United States were more likely to improve their respect for their citizens' rights and freedoms than were countries that did not send their military officers.

8.1 DEFINING AND MEASURING DEMOCRATIC PRACTICES

In the last chapter democracy was defined in terms of a country's political institutions, primarily those political institutions that allow the quintessential democratic act: participation in free and fair elections. A country that has contested and fair elections is a government "of the people" and "by the people." This chapter looks at the third characteristic of Lincoln's classic trilogy, the extent to which a country is also "for the people." In this chapter, democracy is defined in terms of civil and human rights. Democracies are countries in which citizens are permitted individual rights and freedoms. These include the right of all people to be able to go about their daily business without the fear that government officials will apprehend, falsely imprison, torture, or kill them or any of their fellow citizens. Democratic freedoms include the ability to express your opinion in public without fear of reprisal by the government, to travel about freely, to organize, to participate in political movements, and to practice your own religion.

The analyses used Cingranelli-Richards (CIRI) indices that operationalize democracy defined in terms of its practices toward its citizens. The CIRI dataset has the advantage of specifically coding government actions and disaggregating various types of government practices. The CIRI data includes two alternative measures of government respect for citizens' rights. The Physical Integrity Rights Index measures government respect for the physical security of its citizens in the sense that all citizens have the right not to be tortured, murdered, imprisoned for political reasons, and/or made to disappear. In the original CIRI data, the index ranges from 0 (no respect) to 8 (full respect). The other CIRI index, the Empowerment Rights Index, measures government respect for citizens' basic freedoms, which include freedom of movement, speech, religion, political participation, and workers' rights. It ranges from 0 (no respect) to 10 (full respect). For easier interpretation, both indices were reversed so that 0 denoted full respect for rights with increasing values indicating greater abuse of rights.

8.2. MODELING DEMOCRATIC PRACTICES

Modeling the link between a government's propensity to respect the rights and freedoms of its citizens and other events such as coups, democratic transitions, and war has been an important and prominent research agenda in the last several decades. The factors that scholars have identified to have important effects on a country's propensity to abuse the rights and freedoms of its citizens are: its level of economic development, its openness to trade, whether there are ongoing armed conflicts with internal groups, whether there are ongoing armed conflicts with another country, whether it is ruled by a military dictator, whether it was formerly a colony, whether there are population pressures on scare resources, and how long the country has been a democracy. These are the factors included in the models. Data spanned the years 1980–2006 and encompassed 158 countries, each having a population of at least a half million.[1] Political identities were fairly evenly represented within the data with 28% coherent autocracies, 39% coherent democracies, and 33% mixed.[2]

Ordered logit models with random intercepts, a form of multilevel longitudinal models, were used to evaluate the impact of the U.S.-hosted military educational exchanges on the propensity of a country to repress the rights and freedoms of its citizens. Multilevel models were useful and appropriate because the data are longitudinally dependent, clustered (by country), and heavily trended (much the same from year to year). Liang and Zeger (1993) provided a helpful example in order to understand what is meant by longitudinally dependent, clustered, and heavily trended, and why we should care about these data characteristics. They described a study of the development of low birth-weight children. In this study, weight measurements of any one child were dependent on that child's previous weight measurement. Additionally, children within the same family were more likely to be similar to each other than to other children not in their family because the children lived in the same home environment and shared similar genetic material. Because of these data characteristics, the observations made of the children were not statistically independent, but correlated both within families and with past observations.

The structure of the data on democratic practices used for the analyses discussed in this chapter are similar to Liang and Zeger's example of clustered and longitudinally dependent data. Observations are in the form of country-years, meaning the data contains annual observations of the same country over time. Any current values are likely to be correlated with past values, and any one measure of a country is likely to be influenced by factors unique to that country and not others. For example, a country's population density is not statistically independent from other observations, but it depends on how the country's population grew from the previous year and on factors such as birth and death rates, quality of health care, and cultural or economic motivations for a large or small number of children that are unique to that country. One fundamental assumption of regression models, such as the ordered logit/probit models that are often employed in quantitative studies, is that observations are statistically independent. However, when observations are not independent, using standard regression techniques is problematic because it may result in spurious "significant" results and loss of efficiency (Hox 2002; Luke 2004). Multilevel ordered logit models with random intercepts that account for the dependence or correlation between clustered observations avoid these problems.

Random effects models, such as those that were used for the analysis in this chapter, take into account the natural heterogeneity between countries and model country-specific probabilities as well as the overall mean response. The random intercept represents the combined effect of unobserved contextual factors that lead some countries to behave differently from others (Rabe-Hesketh and Skrondal 2008: 247). One might think of it as modeling variations between the genetic makeup of different countries; the genetic makeup is mostly unobservable, but nevertheless it is specific to each country and influences how the country behaves. Models were estimated using adaptive quadrature algorithms to generate the maximum likelihood estimates of the model parameters.[3] Level-1 units were observations for each year for each country (country-years); they were clustered within level-2 units, countries.

Taking into account the findings in the last chapter, each model was generated including a measure for level of democracy and excluding it. The reason for this was to test the extent to which military educational exchanges have a direct impact on democratic practices and the extent to which their impact is indirect. In the last chapter military exchanges were shown to be positively and systematically associated with a country's development of democratic institutions. A large body of research on human rights has also shown that countries with democratic political institutions are less likely to commit human rights abuses than are their autocratic counterparts. Thus, it is necessary to test whether the effects of the military exchanges are on the development of democratic political institutions and if it is these institutions that determine a country's behavior toward its citizens, or whether there is a direct impact of the military educational exchanges on human rights practices.

8.3. RESULTS

Six sets of four models were estimated to assess the impact of military educational exchanges on the propensity of a country to abuse the rights and freedoms of its citizens. To begin, two models using the two different concepts of citizen rights as the dependent variable were estimated. The first assessed the propensity of a country to deny citizens their basic freedoms, as measured by the CIRI Empowerment Rights Index. The second assessed the propensity of a country to falsely imprison, torture, and murder citizens, as measured by the CIRI Physical Integrity Rights Index. In order to test the extent to which military exchanges have an independent and direct effect on democratic practices each of the models was reestimated excluding the measure for a country's level of democracy that was hypothesized to play an intervening role.[4] Finally, each of these four models was estimated using a time lag of one through six years for the key explanatory variable (*Military educational exchanges*) in order to try to capture short-term as well as longer-term impacts of the military exchanges.[5] The results for all twenty-four models are summarized in table 8.1, which provides an overall summary for the key *Military educational exchanges* variable at different time lags. Following a summary of these results, the full models using the three-year lagged variant of the military exchange variable are discussed in more depth.[6]

A quick glance at table 8.1 shows that the results are robust and stable across the different time lags. Negative coefficients indicate that military exchanges had a negative impact on the government's propensity to abuse citizens' rights; that is to say negative coefficients mean that the variable was associated with improving respect for rights. In the top half of table 8.1 it can be seen that military educational exchanges were consistently associated with a decrease in the propensity of a government to deny basic freedoms such as the freedom of movement, speech, religion, political participation, and workers' rights to their citizens. Throughout these analyses, *Military educational exchanges* consistently achieved high levels of significance. In the bottom half of table 8.1, it can be seen that in two of the model variants, military exchanges were significant and associated with a decrease in the propensity of a government to falsely imprison, torture, or murder its citizens. However in all other variants, military exchanges were not significant, indicating that they had no association, either positive or negative, with the propensity of a government to violate its citizens' physical security. To summarize, military educational exchanges had a significant positive and systematic impact on governments' respect for citizens' basic freedoms. But they had very little impact, either positive or negative, on governments' respect for the physical security of their citizens.

The models also allowed the assessment of the extent to which the development of democratic institutions was an intervening factor between military educational

Table 8.1. Military Education Exchange Coefficient in All Models

	Propensity to Deny Basic Freedoms	
lags	*Full Model*	*Excludes* Level *of Democracy*
1	–0.733***	–1.088***
2	–0.563***	–0.863***
3	–0.443***	–0.692***
4	–0.414***	–0.653***
5	–0.313**	–0.499***
6	–0.243*	–0.434***
	Propensity to Falsely Imprison, Torture, or Murder	
lags	*Full Model*	*Excludes* Level *of Democracy*
1	—	–0.412***
2	—	–0.312**
3	—	—
4	—	—
5	—	—
6	—	—

All models were estimated using the twelve variables discussed in the text.
Entries are ordered logit coefficients of the military exchange variable.
*** p<0.01, ** p<.005, * p<0.1
"-" indicates that the coefficient was not significant in that model.

exchanges and a country's democratic practices. In order to assess this relationship, it is necessary to compare the models that included *Level of democratic institutions* with those that excluded it. In the top half of table 8.1, *Military educational exchanges* was significant in all of the models. These results indicate that military exchanges have a direct impact on democratic practices conceived of as citizens' basic freedoms. However, there is also some suggestion that the development of democratic political institutions is a weak intervening factor. By comparing columns 1 and 2 (in the top half of table 8.1), we can see that the coefficient for *Military educational exchanges* consistently became smaller (in absolute terms) when *Level of democratic institutions* was included indicating a weak intervening effect. Nevertheless, the military exchanges were significant whether or not the level of democracy measure was included. To summarize, in the results for democratic practices defined as a country's respect for citizen's basic freedoms, military exchanges have a significant and positive impact both directly and (the finding from the last chapter) because they support the development of democratic institutions, which, in turn, has a positive impact on democratic practices. This finding is robust across all model variants.

Turning now to the bottom half of table 8.1, it can be seen that *Military educational exchanges* was significant in two of the models, but lost significance when *Level of democratic institutions* was included indicating a significant intervening impact. Thus, for democratic practices conceived of as respect for citizens' physical security, military exchanges had no direct impact, either positive or negative, on a country's practices when the level of democratic institutions is taken into account. Any effect that they do have is because (as discussed in the last chapter) countries that participate in the exchanges are more likely to develop democratic institutions than are those countries that do not participate, and this process in turn supports better democratic practices in those countries.

Figures 8.1 and 8.2 show graphical representations of the four models using the three-year lagged *Military educational exchanges* variable in order to discuss in more detail the intervening effect possibly triggered by the *Level of democratic institutions* variable.[7]

As can be seen in figure 8.1, participation in military educational exchanges decreased the propensity of the country to repress citizens' basic freedoms conceived of as basic rights to freedom of movement, speech, religion, political participation, and workers' rights. Countries that sent their military officers to study at U.S. military war and staff colleges were more likely to be associated with improving democratic practices than were countries that did not send their military officers. Predicted probabilities generated for an average hypothetical country showed that participation in military exchanges increased the likelihood of *better* democratic practices by 7.81% in comparison to a country that did not participate and decreased the likelihood of *worse* practices by 6.69%.[8] In comparison with other types of educational exchanges hosted by the United States, the effect is substantial. For the same

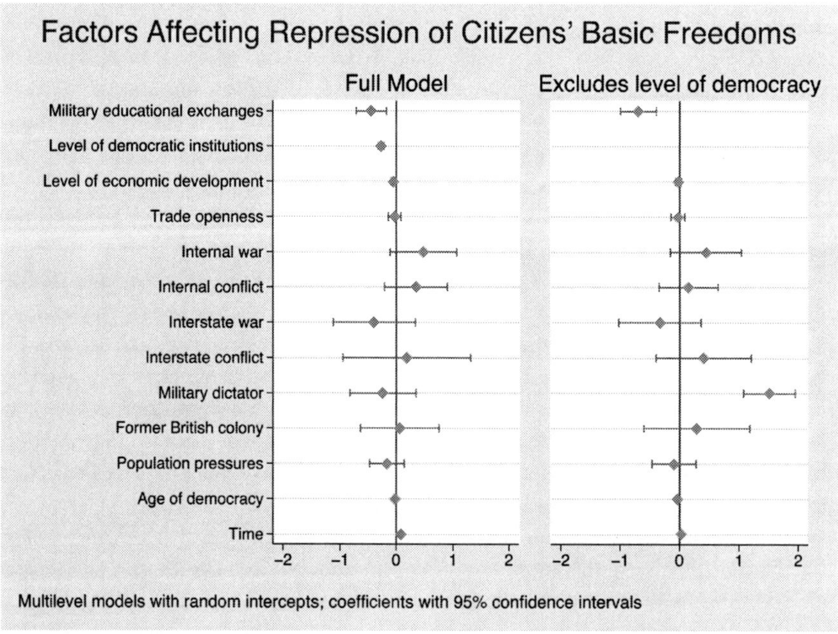

Figure 8.1. Multilevel Model Results, Factors Impacting the Propensity to Deny Basic Freedoms

hypothetical average country, a high level of participation in university-level student exchange programs increased the likelihood of better democratic practices by 2% in comparison to a low level of participation and decreased the likelihood of worse practices by 1.69%.[9]

Turning to the other independent variables, *Level of democratic institutions* was the single best predictor of whether citizens were able to exercise basic freedoms and rights. These results confirmed previous research that found, unsurprisingly, that as democratic institutions flourished, the government was likely to engage in more democratic practices. The significance of *Military dictatorship* when *Level of democratic institutions* was removed from the model indicated that military dictatorships were no worse than civilian dictatorships when it came to their propensity to abuse and repress their citizens.

Figure 8.2 shows the same set of models but with a different measure of democratic practices, the propensity of a government to falsely imprison, torture, or murder citizens.

These two models also used the three-year lagged version of the measure for military educational exchanges. As can be seen, military educational exchanges had no discernible effect on the propensity of a government to physically abuse its citizens. For the other independent variables, the results confirmed findings from previous studies. Government respect for citizens' rights increased as political institutions

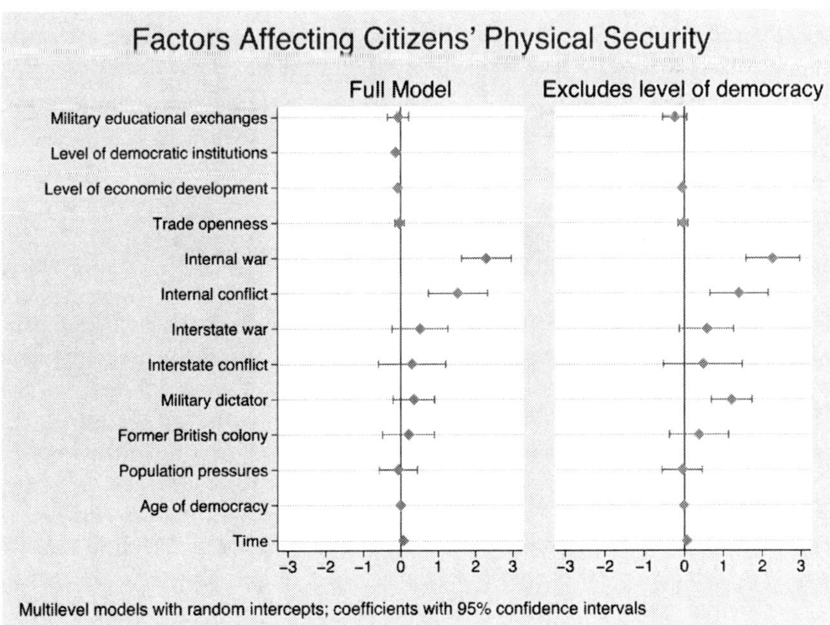

Figure 8.2. Multilevel Model Results, Factors Impacting the Propensity to Falsely Imprison, Torture, or Kill

became more democratic and as the level of economic development rose. Governments were much more likely to repress their citizens when they were involved in ongoing civil wars and other sorts of armed conflicts with internal groups. Once again it can be seen that military dictatorships were no worse than civilian dictatorships when it came to their propensity to abuse and repress their citizens.

It is worth emphasizing that all of the results were obtained while controlling for preconditions that are known to be associated with a government's propensity to respect the rights of its citizens. Additionally, the results were consistent across multiple time lags. While it is always possible to invoke the presence of unobservable factors, the controls in the models represent our best knowledge of the factors affecting government repression of citizens. The inclusion of the random intercepts accounts for the presence of previously unobserved factors as advocated by Przeworski (2007: 161) when he singled out educational exchanges as an important "unobservable" that "may matter." It is also important to recognize that the models do show differential effects: exchanges support the development of respect for basic freedoms and liberties but have no effect on respect for physical integrity. This is important methodologically because it demonstrates that the models do capture disconfirming effects as well as those that support the argument presented in this book.

To summarize, the results show that military educational exchanges do have a positive impact on governments' propensity to respect citizens' rights of movement, speech, religion, political participation, and workers' rights. The results are less promising for the right not to be physically abused. At best, military exchanges may have a beneficial effect, because countries that participate in the military exchanges are more likely to develop democratic political institutions (as was demonstrated in the last chapter), which then leads to a decrease in levels of physical abuse; at worst, military exchanges have no impact on a government's propensity to torture, engage in extrajudicial killing, political imprisonment, and disappearance.

8.4. COUNTRY-SPECIFIC EXAMPLES

Country-specific examples are useful for illustrating the interaction of democratic political institutions, U.S. military educational exchanges, and the propensity of a government to uphold or abuse the rights of its citizens. Figures 8.3, 8.4, and 8.5 illustrate how the model results discussed in this chapter pertain to the specific experiences of different countries. Figure 8.3 shows random intercept predictions with 95% confidence intervals for each country, based on the three-year lagged *Military educational exchanges* model that evaluated respect for basic freedoms (Empowerment Rights Index) and included *Level of democratic institutions*. Negative random intercepts identify countries that had human rights records that were better than expected given the explanatory variables in the model whereas positive random intercepts identify countries that had records worse than expected given the factors modeled. So, for example, countries that had better-than-expected records included

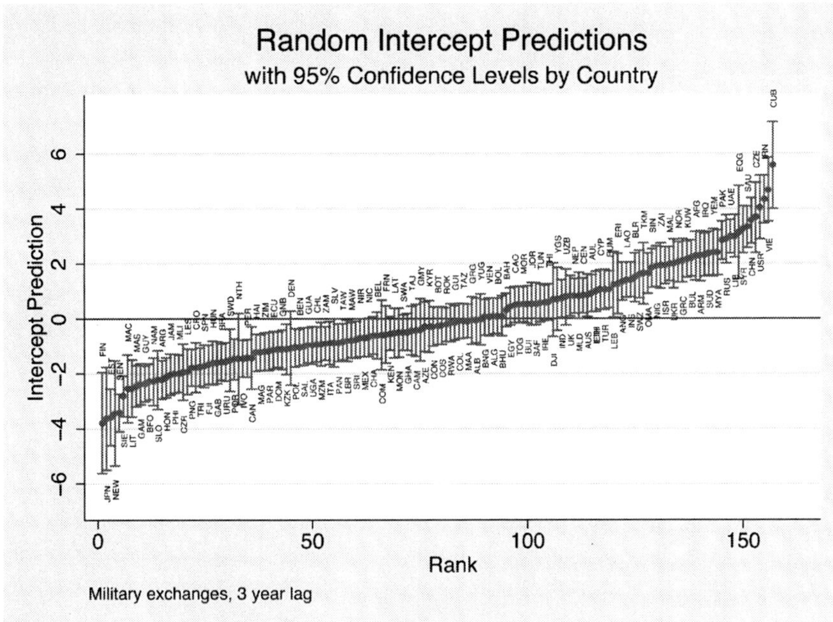

Figure 8.3. Random Intercept Predictions

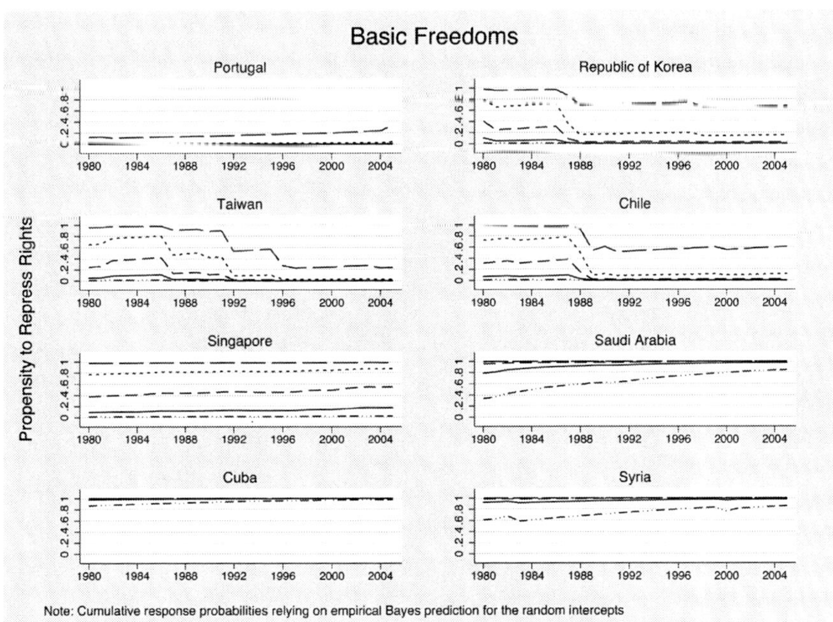

Figure 8.4. Basic Freedoms, The Experience of Individual Countries

Note for each graph: The top line composed of long dashes represents that probability of repression as measured by the CIRI Empowerment Rights Index greater than level 1, Prob (Y > 1); the second from the top line composed of dots represents that probability of physical abuse greater than level 3, Prob (Y > 3); and so forth. The bottom line thus represents the overall propensity of the government to deny citizens the basic rights of freedom of movement, speech, religion, political participation, and workers' rights at its most egregious level.

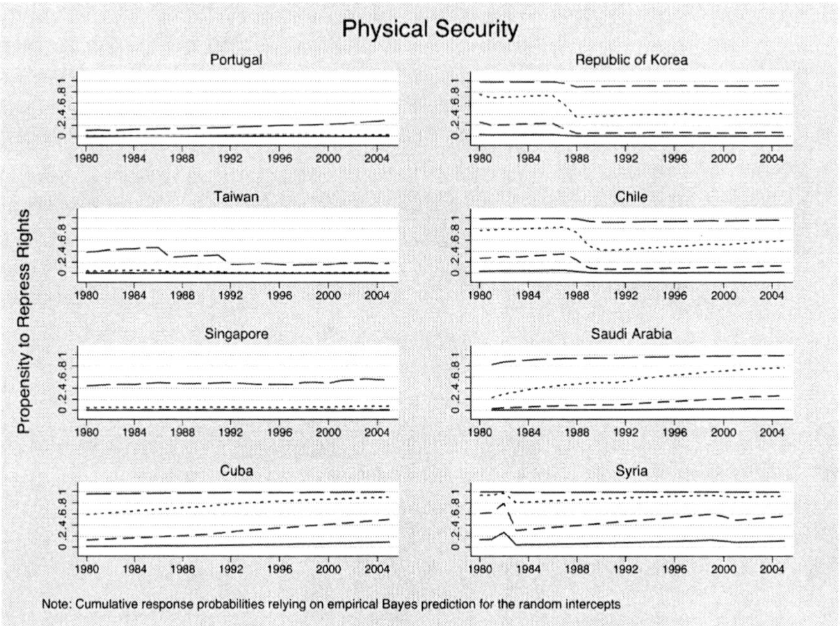

Figure 8.5. Physical Security, The Experience of Individual Countries

Note for each graph: The top line composed of long dashes represents that probability of physical abuse as measured by the CIRI Physical Integrity Rights Index greater than level 1, Prob (Y > 1); the second from the top line composed of dots represents that probability of physical abuse greater than level 3, Prob (Y > 3); and so forth. The bottom line thus represents the overall propensity of the government to falsely imprison, torture, and murder citizens at its most egregious level.

Finland, Japan, Estonia, and New Zealand; countries with worse-than-expected records included Cuba, Vietnam, Iran, and the Soviet Union.

Figures 8.4 and 8.5 each contain eight graphs of country-specific cumulative response probabilities that track the risk to citizens in each country over time.[10] Figure 8.4 shows the estimates of the propensity of the government to repress citizens' basic freedoms, and figure 8.5 shows the estimates of the propensity of the government to harm citizens physically. These graphs illustrate the variety of scenarios encompassed within the statistical analyses.

Starting from the upper left panel of both figures, there is Portugal. Today a democratic country following the Carnation Revolution that overthrew the authoritarian government in 1974, Portugal has consistently sent military officers to U.S. military schools. The risk to citizens, measured as either physical abuse or denial of basic freedoms, is very low throughout the entire period, indicating that democratic freedoms and individual rights were secure for Portuguese citizens. To the right, is the Republic of Korea, a country that transitioned to democracy in 1987 when its first president was elected by popular vote. During both its autocratic and democratic eras, Korea has consistently participated in U.S. military exchanges. Its actions toward its citizens have improved since the late 1980s with a declining propensity to repress citizens' rights. Next is Taiwan, a country that has transitioned to democracy and that has participated in U.S. military educational programs except during the early 1980s. Risk to Taiwanese citizens began to decline before its transition to democracy in 1992. Next to Taiwan, is the graph for Chile, where a military coup in 1973 brought to power the Pinochet military dictatorship that engaged in years of large-scale repression of the worst kind in which thousands of citizens were tortured, imprisoned, executed, murdered, and disappeared. Democracy returned to Chile in late 1989. During most of the years of the military dictatorship, Chilean military officers did not participate in U.S. military educational programs. Repression of the worst sort decreased as democratic institutions were restored, although lesser levels of repression continued. While government respect for citizens' basic freedoms has improved, it can be seen that lower levels of physical abuse continued.

The last four countries are interesting because they have all maintained non-democratic institutions throughout the entire period of the study. The graph for Singapore, an autocratic country that has consistently sent military officers to U.S. military schools, shows that the government's propensity to repress freedoms and threaten the physical security of its citizens has remained at consistently low levels over the years, but with higher levels of government repression of people's basic freedoms. Saudi Arabia is another autocratic country that has consistently sent military officers to U.S. military educational programs; however, the government's tendency to repress and physically abuse citizens has increased over the years. Finally, at the bottom of figures 8.4 and 8.5, there are Cuba and Syria. Both are autocratic countries that have never sent military officers to U.S. military schools and whose governing practices have worsened over the years, in particular basic democratic freedoms are virtually nonexistent and the level of physical abuse has worsened as well.

8.5. SUMMARY

The empirical analyses demonstrate the important longer-term influence that U.S.-hosted military educational exchanges have on the development of not only democratic institutions, but also democratic practices. Countries that sent their military officers to study at military institutes in the United States were more likely to develop better democratic practices, especially in terms of improving their respect for their citizens' basic rights and freedoms, than were countries that did not participate in the exchanges. A country's participation in military exchanges increased the likelihood of *better* democratic practices by almost 8% in comparison to a country that did not participate and decreased the likelihood of *worse* practices by around 7%.

NOTES

1. The first year of observation is 1980 because that was the first year for which CIRI indices were available.

2. Categorization followed standard practice using the Polity IV 21-point scale where 0 is most authoritarian. Highly authoritarian countries were those with a polity rating of 0–3; democratic countries were those with a polity rating of 17–20, and mixed regimes were those in between.

3. For further discussion on multilevel longitudinal models see Liang and Zeger (1993), Luke (2004), and Rabe-Hesketh and Skrondal (2008). Models were estimated using Generalized Linear Latent and Mixed Model (gllamm) estimation commands; Rabe-Hesketh and Skrondal (2008) provide a detailed discussion. I thank Giacomo Chiozza for his advice and assistance on modeling issues.

4. Blalock (1979: 468–77) provides further information on the modeling of intervening variables.

5. All other independent variables were lagged one year. The purpose is to reduce the risk of reverse causation. There was no theoretical reason to extend the multiyear lags to the other independent variables.

6. I chose the three-year lagged variants as representative. Results were consistent across the other versions.

7. The estimated intra-class correlation is the proportion of variance explained by the clustering of units by country. All models had an intra-class correlation of nearly 0.5 or greater, meaning that the expected correlation for any two randomly chosen measurements for any single country was at least 0.5. This indicates that there was a great deal of dependence of observations within countries; and, thus multilevel models were an excellent choice for the analyses.

8. Predicted probabilities were computed by (1) setting the value of the explanatory variables to their median value, (2) setting the value of the random effects to zero, and then (3) switching the dichotomous indicator *Military educational exchanges* from 0 to 1. *Better* practices are defined as a value of 3 or lower; *Worse* practices are defined as a value of 7 or higher on the 11-point dependent variable.

9. For more information comparing these two types of exchange programs see Atkinson (2010).

10. The probabilities rely upon empirical Bayes prediction for the random intercepts.

9

Conclusion and Policy Implications

On June 8th of 2012, a beautiful sunny day, friends and family gathered on Fort Leavenworth's immaculately manicured parade field to watch as more than 1,000 students graduate from their course of military studies at the U.S. Army Command and General Staff College. Each year in early summer, this scene takes place at all of the U.S. military's war and staff colleges. It is a time when the accomplishments of the officers' grueling yearlong course of study are recognized. Amongst the graduates to cross the stage at Fort Leavenworth were sixty-nine international graduates from sixty-five countries. They were the most recent addition to a long line of international graduates stretching back to the first international graduates in the early 1900s. But what happens now that graduation is over and friends and colleagues part ways for new assignments?

As the research in this book shows, after graduation and in the many years to come, the educational experiences at the war and staff colleges have important effects on U.S. national security and across the globe. Military educational exchange programs hosted by the U.S. military's war and staff colleges strengthen U.S. foreign policy because they increase the ability of the U.S. military to operate more effectively with allied and partner nations; they build goodwill toward the United States that is the basis of soft power; and they help the United States to nurture and support democratic cultures in transitioning as well as nondemocratic countries.

In his seminal work on the concept of soft power, Joseph Nye (2004: 8) argued that soft power arises largely "from the values an organization or country expresses in its culture, in the examples it sets by its internal practices and policies, and in the way it handles relations with others." When one has soft power it is easier to achieve goals, as Nye (2004: 5) argued, because others admire your ideas and want to emulate your example. Soft power advocates have often argued that international exchanges could serve as a powerful way to build soft power because exchange participants usually return home with deeper understanding, warm feelings, and personal connections to the people and the country that hosted them. The socialization model and the empirical analyses presented in this book demonstrate how this process unfolds for one important group of people: military officers. The welcoming and integration of military exchange participants into U.S. communities and U.S. military schools, and the personal outcomes these interactions generated show what it means at the level of the individual person to build military soft power. By fostering goodwill, friendships, and social networks, the military educational exchanges are an important component of U.S. public diplomacy.

Survey data showed that the international officers returned home from the United States with deeper understanding and appreciation of the United States, its people, and its values as well as many professional contacts and personal friendships that they hoped to build upon in the coming years. Participation in official classroom activities and off-duty social events with U.S. and other international classmates built common experiences amongst the officers and linked them into the wider epistemic community of military professionals. The community embraced the newcomers and socially integrated both the U.S. officers and their international colleagues through many activities. As the survey results show, making friends with fellow officers, neighbors, and people in their local communities was one of the highlights of the officers' educational experience. Social events with colleagues, traveling around the United States, and family activities and opportunities within local communities for the officers' children facilitated and added depth to the officers' social interactions. All of these contributed to the building of friendships and professional networks as well as to a positive impression of the United States. Through this process the military exchanges helped to build U.S. soft power and served as a source of cross-cultural information for both U.S. and international officers. As an officer from Africa observed, "I learned a lot from my friends and really these experiences will change a lot my future vision."

The resultant military soft power helps the United States achieve one of the stated goals of the United States' International Military Education and Training (IMET) program, which is to "increase the ability of foreign military and civilian personnel to instill and maintain democratic values and protect internationally recognized human rights in their own government and military" (DoD/DoS *Foreign Military Training* 2011: II-2). The empirical analyses showed that the military exchanges were an important element of U.S. foreign policy because those countries that participated were more likely to develop democratic institutions and respect the rights of their citizens than were those countries that did not participate. Countries that participated in the exchanges were more than twice as likely to develop democratic institutions that supported free, fair, and competitive elections. Participation also increased the propensity of the government to respect its citizens' basic freedoms.

Metaphorically, we might liken the military exchanges to fertile soil that provides an environment conducive to the growth and blossoming of government respect for democratic values and practices. Such an environment nurtures all, both long-lasting hardcore dictatorships as well as those countries already on the path to democratic reform. But the point is that without fertile soil, little will grow on parched earth and hard rock. The experiences of the four Arab Spring countries discussed in the opening chapter illustrate the importance of the exchanges in nurturing and developing this "fertile soil"—that is to say, in nurturing and developing the political and military conditions under which peoples' desires for democratic reforms and democratic governance are likely to be fulfilled.

In 2010, before the Arab Spring democratic protests began, the four countries, Tunisia, Egypt, Libya, and Syria, were similar in that they all were ruled by longtime dictators, GDP per capita was relatively low, all were Arab-Muslim countries, and all were thought impervious to a successful democratic uprising of their citizens. Democratization scholars had previously argued that the single most-important determinant of the ability of a country to support a democratic transition was its level of economic development, usually measured as GDP per capita. The GDP per capita in 2010, on the eve of the Arab Spring, for each country was: Tunisia $3,438; Egypt $1,554; Libya $8,705; and Syria $1,942.[1] By looking at these figures alone, an observer would be hard pressed to predict which of the four countries would be likely to throw out its longtime dictator through a popular democratic uprising; as well as to predict in which country the military would support the forces of democracy rather than dictatorship.

External observers might have claimed that in some of these countries dictators had spent a lot of their money on their military, and, thus, the military should be more effective in suppressing protesters in those countries rather than in others. Once again an observer would be hard pressed to argue the case that it was low military expenditures that led to successful democratic transitions, whereas higher military expenditures supported continued autocratic rule. In 2010, on the eve of the Arab Spring, none of the four countries had spent an inordinate amount of money on their military forces; their military expenditures as a percentage of GDP were: Tunisia 1.4%; Egypt 2%; Libya 1.2% (in 2008); and Syria 4.1%. As a comparison, in 2010, Israel spent 6.5% of its GDP on its military, and the United States spent 4.8%.[2] A reason it is difficult to predict these outcomes is that a key factor has been overlooked in studies of democratic transitions: the role that the U.S. military has played in educating and socializing military officers in nondemocratic countries.

To return to the four Arab Spring countries, figure 9.1 shows how much the United States had invested in International Military Education and Training (IMET) funding in each of these countries since 1975.

Although measured in dollars, these dollars represent individual people who were then able to attend education and training courses in U.S. military schools, including the war and staff colleges. As can be seen, the United States military had literally spent decades engaging the military officers and personnel of Tunisia and Egypt in the manner discussed in this book. The militaries in these countries chose to support the path of democratic reform and facilitated the ouster of the country's dictator. On the other hand, the Libyan and Syrian militaries have been isolated from U.S.

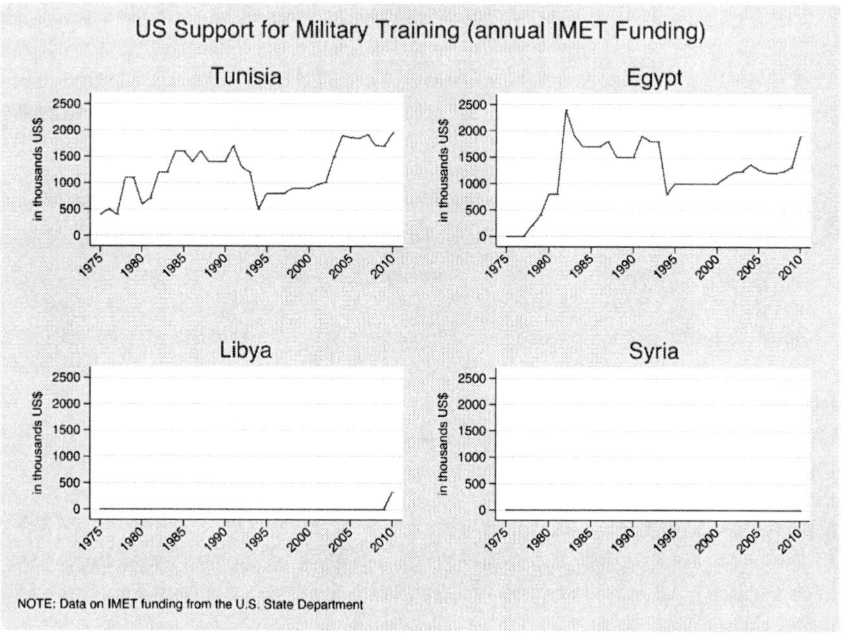

Figure 9.1. U.S. Support for Military Training

military-to-military engagement and abandoned to determine their own destinies. The destiny that they chose was to fight for the regime and against democratic reform. If, *ex ante,* an observer had known of the sustained efforts that the U.S. military had made to build social and professional networks in each of these countries and the impacts these efforts were likely to have based on the findings in this book, then they might not have been so surprised by the success of democratic protests in Tunisia and Egypt.

Does this mean that military educational exchanges "cause" democracy? Certainly not. As mentioned in the introduction, this is no more the case than saying economic development "causes" democracy or military expenditures "cause" autocracy. But what it does mean is that when people who are repressed by autocratic regimes decide to take matters into their own hands by protesting and demanding democratic reforms, they are more likely to succeed because their own militaries are less likely to help the repressive regime stay in power. In both Tunisia and Egypt the military took an active role in facilitating the transition. This was also the case several years earlier in Indonesia, where President Susilo Bambang Yudhoyono was widely credited with bringing peaceful democratic change to his country. Former U.S. Secretary of State Condoleezza Rice (2011: 448) has called Yudhoyono, a former military officer and 1991 alumni of the U.S. Army Command and General Staff College, the "personification of the success of our International Military Education and Training program" who brought Indonesia "back from the brink."

In 2005, Yudhoyono became the first sitting head of state to be inducted into the U.S. Army Command and General Staff College's International Hall of Fame. Reflecting many of the themes that were discussed in this book, in his remarks at this ceremony Yudhoyono talked about how the U.S. military war and staff college exchanges affected their participants; he said the experience

> allows us to learn about each other, and from one another as soldiers, as nations, as citizens of the world. I learned from my peers here that honor and duty crosses national boundaries. We learned from each other the value of democracy, of human rights, as well as the importance of sovereignty. We learned from each other how to move forward in this tenuous obstacle course called nation-building. That immeasurable experience was strengthened by the friendships made during my stay. Friendships, between nations as well as individuals, are what relieve us in times of crisis. (Yudhoyono 2005)

For U.S. Congressional leaders the decision to support and fund military educational exchanges is critical. As the findings in this book show, for a very small sum of money the exchanges at the U.S. military's war and staff colleges build a tremendous amount of goodwill among a critically important institution in other countries: the military. In the past the U.S. government has used military exchanges as a negative sanction; prohibiting or limiting attendance by countries with poor human rights records. Condoleezza Rice (2011: 448–49) argued that, while Congress may be able to send an important signal with these actions, these types of sanctions "have a downside, as the United States loses contact and influence" within those militaries. The *Wall Street Journal* (2009) was more blunt and called such actions "not so 'smart power'" and lamented that when the Pressler Amendment of the 1990s barred U.S. military-to-military contacts with Pakistan, "This only reduced U.S. sway in Islamabad and meant that a generation of Pakistani officers had little contact with the U.S. in their formative years, coloring their views today."

The analysis in the preceding chapters reinforces these statements by showing how military educational exchanges build U.S. soft power and over the long term support those who seek to throw off dictatorships. Engagement gives the United States the ability to build lasting and influential relationships, whereas isolating the militaries in autocratic countries does not provide the opportunity for the militaries in those countries to learn how to choose a different destiny. The exchanges hosted by the U.S. military's war and staff colleges are a proven way that the United States can engage in "wars of ideas," build its own soft power, and promote democracy, soldier by soldier.

NOTES

1. Based on GDP data from the U.S. Department of Agriculture's Economic Research Service's *International Macroeconomic Data Set*; see USDA/ERS (2012).

2. Military expenditure data is from the *SIPRI Military Expenditures Database*; see Stockholm International Peace Research Institute (2012). The last year of data for Libya was 2008.

References

Adler, Emanuel. 1992. "The Emergence of Cooperation: National Epistemic Communities and the International Evolution of the Idea of Nuclear Arms Control." *International Organization* 46:1, 101–45.

Air Command and Staff College (ACSC GOE). 2012. *Gathering of Eagles*. Online at: http://www.goefoundation.com/Brochure%20-%20AY12.pdf; accessed on 8 February 2012.

Air University International Officer School (AU IOS). 2012. *International Honor Roll*. Online at: http://www.au.af.mil/au/spaatz/ios/honor roll.htm; accessed on 18 June 2012.

Air University Spaatz Center (AU Spaatz). 2010. Spaatz Center Homepage. Online at: http://www.au.af.mil/au/spaatz/; accessed on 24 September 2010.

Air War College (AWC Demographics). 2012. *AWC Student Demographics, Class of 2012*. Online at: http://www.au.af.mil/au/awc/students.htm; accessed on 8 February 2012.

Alderson, Kai. 2001. "Making Sense of State Socialization." *Review of International Studies* 27:3, 415–33.

Arbatov, Georgi. 1992. *The System: An Insider's Life in Soviet Politics*. New York: Random House.

Associated Press. 2010. "U.S. Army War College Carlisle Welcomes New Class." pennlive.com, 26 June 2010. Online at: http://www.pennlive.com/midstate/index.ssf/2010/06/us_army_war_college_carlisle.html; accessed on 22 September 2010.

Atkinson, Carol. 2006. "Constructivist Implications of Material Power: Military Engagement and the Socialization of States 1972–2000." *International Studies Quarterly* 50:3, 509–37.

Atkinson, Carol. 2010. "Does Soft Power Matter? A Comparative Analysis of Student Exchange Programs 1980–2006." *Foreign Policy Analysis* 6:1, 1–22.

Babbie, Earl. 2007. *The Practice of Social Research*. 11th Edition. Belmont, CA: Thomson Higher Education.

Baldwin, David A. (Ed.) 1993. *Neorealism and Neoliberalism: The Contemporary Debate*. New York: Columbia University Press.

Ball, Harry P. 1983. *Of Responsible Command: A History of the U.S. Army War College*. Carlisle Barracks, PA: Alumni Association of the United States Army War College.

Baltos, Corey. 2010. "USAWC Graduates Senior Military and Government Leaders." Army War College Public Affairs. Online at: http://www.army.mil/-news/2010/06/14/40758-usawc-graduates-senior-military-and-government-leaders/; accessed on 23 September 2010.

Barnett, Michael and Raymond Duvall. 2005. "Power in International Relations." *International Organization* 59:1, 39–75.

Baruch, Yehuda. 1999. "Response Rate in Academic Studies—A Comparative Analysis." *Human Relations* 52:4, 421–38.

Berger, Peter L. and Thomas Luckmann. 1966. *The Social Construction of Reality: A Treatise on the Sociology of Knowledge*. New York: Anchor Books.

Blair, Jason. 2002. "Marine Corps University Students Graduate." Official Homepage of the US Marine Corps. Online at: http://www.marines.mil/unit/mcbquantico/Pages/2002/Marine%20Corps%20University%20students%20graduate.aspx; accessed on 21 September 2010.

Blalock, Hubert M., Jr. 1979. *Social Statistics*. New York: McGraw-Hill.

Box-Steffensmeier, Janet M. and Bradford S. Jones. 2004. *Event History Modeling: A Guide for Social Scientists*. New York: Cambridge University Press.

Box-Steffensmeier, Janet M. and Christopher J. W. Zorn. 2001. "Duration Models and Proportional Hazards in Political Science." *American Journal of Political Science* 45:4, 972–88.

Brown, Edward, USAF Colonel and Commandant of the International Officer School. 2011. Interview conducted on 8 December 2011, Maxwell Air Force Base, Alabama.

Bull, Hedley. 1977. *The Anarchical Society: A Study of Order in World Politics*. New York: Columbia University Press.

Bush, George W. 2010. *Decision Points*. New York: Crown Publishers.

Checkel, Jeffrey T. 2001. "Why Comply? Social Learning and European Identity." *International Organization* 55:3, 553–88.

Chiozza, Giacomo. 2009. *Anti-Americanism and the American World Order*. Baltimore, MD: Johns Hopkins University Press.

Cingranelli, David L. and David L. Richards. 2008a. *The Cingranelli-Richards (CIRI) Human Rights Dataset*, version 2008.03.12. Online at: http://www.humanrightsdata.org; accessed on 13 January 2009.

Cingranelli, David L. and David L. Richards. 2008b. *The Cingranelli–Richards (CIRI) Human Rights Data Project Coding Manual*, version 2008.3.13. Online at: http://ciri.binghamton.edu/documentation/ciri_coding_guide.pdf; accessed on 13 January 2009.

Clausewitz, Carl von. 1832. *On War*. Edited and translated by Michael Howard and Peter Paret. Princeton, NJ: Princeton University Press, 1976.

Clinton, Hillary Rodham. 2010. "Leading through Civilian Power: Redefining American Diplomacy and Development." *Foreign Affairs* 89:6, 13–24.

Cohen, Gideon. 2005. "Propensity-Score Matching and the Lenin School." *Journal of Interdisciplinary History* 36:2, 209–32.

College of International Security Affairs (CISA). 2012. *Mission*. Online at: http://www.ndu.edu/cisa/about_mission.cfm; accessed on 12 April 2012.

Cook, Steven A. 2011. "Tunisia after Ben Ali." *First Take*. Council on Foreign Relations. Online at: www.cfr.org/democracy-and-human-rights/tunisia-after-ben-ali/p23804; accessed on 2 May 2011.

Cope, John A. 1995. *International Military Education and Training: An Assessment*. Washington, DC: National Defense University, Institute for National Strategic Studies.

Cress, Joseph. 2011. "U.S. Army War College Class of 2012 Reports for Duty." *The Sentinel*, 7 August 2011. Online at: http://cumberlink.com/news/local/military/article_c3a0b504-c15c-11e0-8ba5-001cc4c03286.html; accessed on 6 February 2012.

Dawson, Richard E. and Kenneth Prewitt. 1969. *Political Socialization*. Boston, MA: Little, Brown and Company.

Deichert, Kelly. 2011. "Air War College Graduates Make Impression." *Inside Maxwell AFB*, 26 May 2011. Online at: http://www.maxwell.af.mil/news/story.asp?id=123257776; accessed on 9 February 2012.

Doyle, Michael W. 1983a. "Kant, Liberal Legacies, and Foreign Affairs." *Philosophy and Public Affairs* 12:3, 205–35.

Doyle, Michael W. 1983b. "Kant, Liberal Legacies, and Foreign Affairs, Part 2." *Philosophy and Public Affairs* 12:4, 323–53.

Doyle, Michael. 1986. "Liberalism and World Politics." *American Political Science Review* 80:4, 1151–69.

Drew, Christopher. 2011. "Costly Drone Is Poised to Replace U-2 Spy Plane." *New York Times*, 2 August 2011. Online at: http://www.nytimes.com/2011/08/03/business/global-hawk-is-poised-to-replace-u-2-spy-plane.html?pagewanted=all; accessed on 6 February 2012.

Dunphy, Christopher J. 2000. "CGSC International." *Soldiers*, October 2000, 30–33.

Economist. 2010. *Democracy Index 2010*. New York: The Economist Intelligence Unit Limited.

Feaver, Peter. 1999. "Civil-Military Relations." *Annual Review of Political Science* 2, 211–41.

Finnemore, Martha. 1996. *National Interests in International Society*. Ithaca, NY: Cornell University Press.

Fort Leavenworth Public Affairs Office (PAO). 2011. *Fort Leavenworth Statistics, as of 30 September 2011*. Online at: http://garrison.leavenworth.army.mil/Garrison/media/Garrison-Media-Library/PAO-FtLVN-FY11-Statistics.pdf; accessed on 13 February 2012.

Fort Leavenworth Public Affairs Office (PAO). 2010. *Fort Leavenworth Statistics, as of 30 September 2010*. Online at: http://garrison.leavenworth.army.mil/Garrison/media/Garrison-Media-Library/PAO-FtLVN-FY10-Statistics.pdf; accessed on 13 February 2012.

Fort Leavenworth Public Affairs Office (PAO). 2009. *Fort Leavenworth Statistics, as of 30 September 2009*. Online at: http://garrison.leavenworth.army.mil/Garrison/media/Garrison-Media-Library/PAO-FtLVN-FY09-Statistics.pdf; accessed on 13 February 2012.

George, Alexander L. 1969. "The 'Operational Code:' A Neglected Approach to the Study of Political Leaders and Decision-Making." *International Studies Quarterly* 13:2, 190–222.

Gheciu, Alexandra. 2005. "Security Institutions as Agents of Socialization? NATO and the 'New Europe.'" *International Organization* 59:4, 973–1012.

Gilmore, Gerry J. 2006. "Anti-Terror War Is Struggle of Ideas, Vice Chairman Says." 29 March 2006. Online at: http://www.defenselink.mil/news/newsarticle.aspx?id=15016; accessed on 27 June 2010.

Gilpin, Robert. 1981. *War and Change in World Politics*. New York: Cambridge University Press.

Greater Kansas City Chapter of People to People International (GKC People to People). 2010. *International Military Student Sponsorship*. Online at: www.gkcptp.org/international_officers; accessed on 7 October 2010.

Gregg, Becca. 2011. "Danish Officer Inducted into Army War College Hall of Fame." *Sentinel Reporter*, 24 January 2011. Online at: http://cumberlink.com/news/local/military/article_5cb84760-27e9-11e0-ab47-001cc4c03286.html; accessed on 6 February 2012.

Grieco, Joseph M. 1988. "Anarchy and the Limits of Cooperation: A Realist Critique of the Newest Liberal Institutionalism." *International Organization* 42:3, 485–507.

Haas, Peter M. 1992. "Introduction: Epistemic Communities and International Policy Coordination." *International Organization* 46:1, 1–35.

Hafner-Burton, Emilie M. 2012. "International Regimes for Human Rights." *Annual Review of Political Science* 15, 265–86.

Hafner-Burton, Emilie M. 2005. "Trading Human Rights: How Preferential Trade Agreements Influence Government Repression." *International Organization* 59:3, 593–629.

Hamid, Shadi. 2011. "The Struggle for Middle East Democracy." *Cairo Review of Global Affairs*.

Hastings, Max. 1987. *The Korean War*. New York: Simon & Schuster.

Hepburn, Mary A. (Ed.) 1995. "Revitalizing Political Socialization Research: An Introduction to the Symposium." *Perspectives on Political Science* 24:1, 5–6.

Hickey, James R. 2001. *The History and Significance of the U.S. Army War College International Fellows Program*. Carlisle Barracks, PA: U.S. Army War College.

Holsti, Ole R. 1967. "Cognitive Dynamics and Images of the Enemy: Dulles and Russia." In David J. Finlay, Richard R. Fagen, and Ole R. Holsti (Eds.), *Enemies in Politics*. Chicago, IL: Rand McNally, 25–96.

Homan, Michael, Dean of Students at the International Officer School. 2011. Interview conducted on 8 December 2011, Maxwell Air Force Base, Alabama.

Hosmer, David W., Jr., and Stanley Lemeshow. 1999. *Applied Survival Analysis: Regression Modeling of Time to Event Data*. New York: John Wiley & Sons Inc.

Howard, Michael. 2004. *War and the Liberal Conscience*. 5th Edition. New Brunswick, NJ: Rutgers University Press.

Hox, Joop. 2002. *Multilevel Analysis: Techniques and Applications*. Mahwah, NJ: Lawrence Erlbaum Associates.

Ikenberry, G. John and Charles A. Kupchan. 1990. "Socialization and Hegemonic Power." *International Organization* 44:3, 283–315.

Industrial College of the Armed Forces (ICAF). 2010. *ICAF History*. Online at: www.ndu.edu/icaf/about/history.htm; accessed on 30 September 2010.

Institute of International Education (IIE). 2011. *A Brief History of IIE*. Online at: www.iie.org/Who-We-Are/History; accessed on 25 October 2011.

Jervis, Robert. 1976. *Perception and Misperception in International Politics*. Princeton, NJ: Princeton University Press.

Johnson, Dominic D. P. and Dominic Tierney. 2006. *Failing to Win: Perceptions of Victory and Defeat in International Politics*. Cambridge, MA: Harvard University Press.

Johnston, Alastair Iain. 2001. "Treating International Institutions as Social Environments." *International Studies Quarterly* 45:4, 487–515.

Jordan, Amos A., William J. Taylor, Jr., and Michael J. Mazarr. 1999. *American National Security*. Baltimore, MD: Johns Hopkins University Press.

Kamrava, Mehran. 2000. "Military Professionalization and Civil-Military Relations in the Middle East." *Political Science Quarterly* 115:1, 67–92.

Kant, Immanuel. 1795. *Perpetual Peace: A Philosophical Sketch*.

Kennedy, Robby. 2003. "International Officers Visit Kansas Capitol." *Fort Leavenworth Lamp*, 7 August 2003, 14.

Keohane, Robert O. 1984. *After Hegemony: Cooperation and Discord in the World Political Economy*. Princeton, NJ: Princeton University Press.

Kerry, John. 2013. Address at the University of Virginia, Charlottesville, VA, 20 February 2013. Online at: www.state.gov/secretary/remarks/2013/02/205021.htm, accessed on 28 March 2013.

Kinsman, Jeremy. 2011. "Democracy Rising: Tunisia and Egypt, When Idealists Got It Right." *Policy Options*, April 2011, 37–43. Montréal: Institute for Research on Public Policy.

Kirkpatrick, David D. 2011a. "Military Backs New Leaders in Tunisia." *New York Times*, 16 January 2011.

Kirkpatrick, David D. 2011b. "Tunisia Names Unity Government." *New York Times*, 17 January 2011.

Kok, Andre. 2010. "Officials Select Potential Base for Royal Saudi Air Force F-15 Training Unit." *Air Force News*. Online at: http://www.af.mil/news/story.asp?storyID=123234557; accessed on 12 December 2010.

Kratzer, Christopher. 2011. "ACSC Students Set High Standard." *Inside Maxwell AFB*, 10 June 2011. Online at: http://www.maxwell.af.mil/news/story.asp?id=123259543; accessed on 9 February 2012.

Lawson, George. 2008. "For a Public International Relations." *International Political Sociology* 2:1, 17–37.

Lebow, Richard Ned. 2009. *A Cultural Theory of International Relations*. New York: Cambridge University Press.

Liang, Kung-Yee and Scott L. Zeger. 1993. "Regression Analysis for Correlated Data." *Annual Review of Public Health* 14, 43–68.

Liewer, Steve. 2010. "Army Staff College Graduates 985." Homepage of the United States Army. Online at: www.army.mil/-news/2010/06/11/40736-army-staff-college-graduates-985/; accessed on 23 September 2010.

Lipset, Seymour Martin. 1996. *American Exceptionalism: A Double-Edged Sword*. New York: W. W. Norton & Company.

Lugo, Adrian A. 2003. "International Relations: Officers Learn about Media, 1st Amendment during Visit to University of Kansas School of Journalism." *Fort Leavenworth Lamp*, 20 February 2003, 12.

Luke, Douglas A. 2004. *Multilevel Modeling*. Thousand Oaks, CA: Sage.

Lutterbeck, Derek. 2012. "Arab Uprisings, Armed Forces, and Civil-Military Relations." *Armed Forces and Society* (online 13 April 2012), 1–25.

LV Mag. 2009. "Cultural Exchange: The Volunteers of Operation International." *LV Mag: Life in Leavenworth County*, February/March 2009, 25–30.

Manning, David F. 2010. *Global Arms of Collegiality*. Lawrence, KS: CreateSpace Inependent Publishing Platform.

March, James G. and Johan P. Olsen. 1998. "The Institutional Dynamics of International Political Orders." *International Organization* 52:4, 943–69.

Marine Corps University (MCU). 2012. *Marine Corps University Factbook 2011–2012*. Quantico, VA: MCU Institutional Research and Assessment & Planning Marine Corps University Press.

Marine Corps University Foundation (MCU Foundation). 2010. *MCU History*. Online at: http://www.mcuf.org/mcu_history.html; accessed on 24 September 2010.

Marshall, Monty G., Keith Jaggers, and Ted Robert Gurr (Marshall et al., *Data*). 2011. *Polity IV Project: Political Regime Characteristics, 1800–2010*. Center for Systemic Peace. Online at: http://www.systemicpeace.org/polity/polity4.htm; accessed on 25 February 2012.

Marshall, Monty G., Keith Jaggers, and Ted Robert Gurr (Marshall et al., *Users' Manual*). 2011. *Polity IV Project: Political Regime Characteristics and Transitions, 1800–2010, Dataset Users' Manual*. Online at: http://www.systemicpeace.org/inscr/p4manualv2010.pdf; accessed on 25 February 2012.

Mattis, James N. 2011. *Statement of General James N. Mattis, U.S. Marine Corps Commander U.S. Central Command before the Senate Armed Services Committee on the Posture of U.S. Central Command.* 1 March 2011. Online at: http://armed-services.senate.gov/statemnt/2011/03%20March/Mattis%2003-01-11.pdf; accessed 9 February 2012.

Maxwell Air Force Base. 2009. "Class Begins for Air War College." *Inside Maxwell AFB*, 7 August 2009. Online at: http://www.maxwell.af.mil/news/story.asp?id=123162348; accessed on 24 September 2010.

McCarthy, George, Jr., Deputy Director for Air University International Affairs and Education. 2011. Interview conducted on 8 December 2011, Maxwell Air Force Base, Alabama.

McKinley, James. C., Jr. 2008. "Fidel Castro Resigns as Cuba's President." *New York Times*, 20 February 2008.

McLeod, Jack M. and Dhavan V. Shah. 2009. "Communication and Political Socialization: Challenges and Opportunities for Research." *Political Communication* 26:1, 1–10.

Mead, Margaret. 1942. *And Keep Your Powder Dry: An Anthropologist Looks at America.* New York: William Morrow & Co.

Mearsheimer, John J. 1994–1995. "The False Promise of International Institutions." *International Security* 19:3, 5–49.

Miller, Arthur H. 2006. "Promoting Democratic Values in Transitional Societies through Foreign Aid." Paper presented at the Midwest Political Science Association Annual Meeting, Chicago, IL.

Mueller, John. 2004. *The Remnants of War.* Ithaca, NY: Cornell University Press.

Munck, Gerardo L. and Jay Verkuilen. 2002. "Conceptualizing and Measuring Democracy." *Comparative Political Studies* 35:1, 5–34.

National Defense University (NDU). 2012. *International Fellows Hall of Fame.* Online at: http://www.ndu.edu/info/IFHall_of_fame.cfm; accessed on 13 February 2012.

National Defense University (NDU *History*). 2012. *History of the National War College.* Online at: www.ndu.edu/nwc/index.htm; accessed on 12 April 2012.

National Defense University (NDU *Factsheet*). 2010. *Factsheet: AY 2009–2010.* Online at: http://www.ndu.edu/info/NDU%20Factsheet.pdf; accessed on 19 February 2012.

National Defense University (NDU *Graduation*). 2010. *National Defense University Graduation to be Held June 10th.* Press release No. 10-003, May 20, 2010. Washington, DC: National Defense University, Director of Public Affairs. Online at: www.ndu.edu/info/PressReleasesPDF/Graduation_Press_Release_2010.pdf; accessed on 21 September 2010.

National Defense University (NDU *History*). 2010. *History of the National War College.* Online at: www.ndu.edu/nwc/index.htm; accessed on 21 September 2010.

National Defense University, International Student Management Office (NDU ISMO). 2012. *International Fellows Hall of Fame Program.* Online at: http://www.ndu.edu/ismo/index.cfm?secID=169&pageID=1-2&type=section; accessed on 13 February 2012.

National Defense University, International Student Management Office (NDU ISMO *History of IF Program). 2012. History of the International Fellows Program at NDU.* Online at: http://www.ndu.edu/ismo/index.cfm?secID=189&pageID=98&type=section; accessed on 12 April 2012.

Niemi, Richard G. and Mary A. Hepburn. 1995. "The Rebirth of Political Socialization." *Perspectives on Political Science* 24:1, 7–16.

Nye, Joseph S., Jr. 2011. *The Future of Power.* New York: PublicAffairs.

Nye, Joseph S., Jr. 2004. *Soft Power: The Means to Success in World Politics.* New York: Public Affairs.

Nye, Joseph S., Jr., and William A. Owens. 1996. "America's Information Edge." *Foreign Affairs* 75:2, 20–36.

Organization for Security and Co-operation in Europe (OSCE). 1990. *Charter of Paris for a New Europe.* Paris: Conference on Security and Co-operation in Europe.

Pevehouse, Jon C. 2005. *Democracy from Above: Regional Organizations and Democratization.* New York: Cambridge University Press.

Pevehouse, Jon C. 2002. "Democracy from the Outside-In? International Organizations and Democratization." *International Organization* 56:3, 515–49.

Pitkin, Hanna Fenichel. 1972. *Wittgenstein and Justice.* Berkeley, CA: University of California Press.

Przeworski, Adam. 2007. "Is the Science of Comparative Politics Possible?" In Carles Boix and Susan C. Stokes (Eds.), *Oxford Handbook of Comparative Politics.* New York: Oxford University Press, 147–71.

Rabe-Hesketh, Sophia and Anders Skrondal. 2008. *Multilevel and Longitudinal Modeling Using Stata.* 2nd Edition. College Station, TX: Stata Press.

Ralston, Joseph W. 2010. Interview conducted on 18 March 2010.

Reichley, John. 1994. *International Officers: A Century of Participation at the United States Army Command and General Staff College.* Fort Leavenworth, KS: U.S. Army Command and General Staff College.

Reveron, Derek S. 2010. *Exporting Security: International Engagement, Security Cooperation, and the Changing Face of the U.S. Military.* Washington, DC.: Georgetown University Press.

Rice, Condoleezza. 2011. *No Higher Honor: A Memoir of My Years in Washington.* New York: Crown Publishers.

Rice, Condoleezza. 2006. "Remarks at the U.S. University Presidents Summit on International Education Dinner," Washington, DC, 5 January 2006.

Richmeier, John. 2011. "Ceremony Kicks Off New Class at Fort." *Leavenworth Times*, 9 August 2011. Online at http://www.leavenworthtimes.com/news/x1852602759/Ceremony-kicks-off-new-class-at-fort; accessed on 6 February 2012.

Richmeier, John. 2010. "Two Inducted into IHOF." *Leavenworth Times*, 9 April 2010, 1.

Richmond, Yale. 2003. *Cultural Exchange and the Cold War: Raising the Iron Curtain.* University Park, PA: Pennsylvania State University Press.

Ricks, Thomas. 2009. *The Gamble: General David Petraeus and the American Military Adventure in Iraq, 2006–2008.* New York: Penguin Press.

Rosecrance, Richard. 1986. *The Rise of the Trading State: Commerce and Conquest in the Modern World.* New York: Basic Books.

Rummel, R. J. 1983. "Libertarianism and International Violence." *Journal of Conflict Resolution* 27:1, 27–71.

Rumsfeld, Donald. 2006. "Rumsfeld: U.S. Losing War of Ideas: Defense Secretary Says U.S. Fares Poorly in Efforts to Counter Terrorism." Comments by Donald Rumsfeld at the U.S. Army War College, Carlisle, Pa., 27 March 2006. Online at: www.cbsnews.com/stories/2006/03/27/terror/main1442811.shtml; accessed on 10 May 2008.

Sapiro, Virginia. 2004. "Not Your Parents' Political Socialization: Introduction for a New Generation." *Annual Review of Political Science* 7, 1–23.

Sarles, Harrison. 2012. Public Affairs Officer for the U.S. Army Command and General Staff College. Communication with the author, 17 February 2012.

Schelling, Thomas. 1966. *Arms and Influence.* New Haven, CT: Yale University Press.

Schelling, Thomas. 1960. *The Strategy of Conflict.* Cambridge, MA: Harvard University Press.

Schimmelfennig, Frank. 2000. "International Socialization in the New Europe: Rational Action in an Institutional Environment." *European Journal of International Relations* 6:1, 109–39.

Selltiz, Claire, June R. Christ, Joan Havel, and Stuart W. Cook. 1963. *Attitudes and Social Relations of Foreign Students in the United States*. Minneapolis, MN: University of Minnesota Press.

Seltzer, Rick. 2009. "U.S. Army War College Inducts Canadian General into International Fellows Hall of Fame." *The Patriot-News*, 23 September 2009. Online at: http://www.pennlive.com/midstate/index.ssf/2009/09/us_army_war_college_inducts_ca.html; accessed on 5 October 2010.

Sigel, Roberta. 1965. "Assumptions about the Learning of Political Values." *Annals of the American Academy of Political and Social Science* 361, 1–9.

Small, Melvin and David J. Singer. 1976. "The War Proneness of Democratic Regimes 1816–1965." *Jerusalem Journal of International Relations* 1:4, 50–69.

Sorenson, David. S. Forthcoming. *Civil-Military Relations in the Middle East: A Comparative Analysis*. Ann Arbor, MI: University of Michigan Press.

Sorenson, David S. 2010. "Civil-Military Relations." In David S. Sorenson (Ed.), *Interpreting Middle East: Essential Themes*. Boulder, CO: Westview Press, 125–55.

Stattel, Victoria A. 2010. *The Nights in Kabul Are Beautiful*. American Women Veterans Welcome Home Essay. Online at: http://americanwomenveterans.org/home/2010/11/the-nights-in-kabul-are-beautiful/; accessed on 11 November 2010.

Steele, Robert (Major General). 2010. "Commandant's Comments." *From the Rotunda*, Winter/Spring, 1–2. Washington, DC: National War College Alumni Association Newsletter.

Stewart, Michael David. 2010. *Raising a Pragmatic Army: Officer Education at the U.S. Army Command and General Staff College, 1946–1986*. PhD Dissertation, University of Kansas, Department of History, approved on 22 April 2010.

Stockholm International Peace Research Institute (SIPRI). 2012. *SIPRI Military Expenditures Database*. Online at http://milexdata.sipri.org. The data were downloaded on 7 June 2012.

Strange, Susan. 1982. "Cave! Hic Dragones: A Critique of Regime Analysis." *International Organization* 36:2, 479–93.

Summers, Harry G. 1982. *On Strategy: A Critical Analysis of the Vietnam War*. Novato, CA: Presidio Press.

Sun Tzu. 1963. *The Art of War*. Translated by Samuel B. Griffith. London: Oxford University Press.

Sunal, Dennis W., and Cynthia C. Sunal. 1991. "Professional and Personal Effects of the American Fulbright Experience in Africa." *African Studies Review* 34:2, 97–123.

Tannenwald, Nina. 2008. *The Nuclear Taboo: The United States and the Non-Use of Nuclear Weapons since 1945*. New York: Cambridge University Press.

Therneau, Terry M. and Patricia M. Grambsch. 2000. *Modeling Survival Data: Extending the Cox Model*. New York: Springer-Verlag.

Thies, Cameron G. 2012. "International Socialization Processes vs. Israeli National Role Conceptions: Can Role Theory Integrate IR Theory and Foreign Policy Analysis?" *Foreign Policy Analysis* 8:1, 25–46.

Thies, Cameron G. 2003. "Sense and Sensibility in the Study of State Socialisation: A Reply to Kai Alderson." *Review of International Studies* 29:4, 543–50.

Thies, Cameron G. 2001. "A Social Psychological Approach to Enduring Rivalries." *Political Psychology* 22:4, 693–725.

United Nations (UN). 2000. *Demographic Yearbook CD-ROM, Historical Supplement 1948–1997*. United Nations, Statistics Division.

United Nations (UN). 2009. *GDP per capita*. Online at: http://data.un.org/; accessed on 14–15 January 2009.

U.S. Agency for International Development (USAID). 2010. *U.S. Overseas Loans and Grants, Foreign Assistance Fast Facts: FY2010*. Online at: http://gbk.eads.usaidallnet.gov/; accessed on 6 February 2012 and 10 May 2012.

U.S. Air Force, 49th Fighter Wing. 2010. *German Air Force Fact Sheet*. Online at: http://www.holloman.af.mil/library/factsheets/factsheet.asp?id=4364; accessed on 18 November 2010.

U.S. Army Air Forces (USAAF). 1947. *Training of Foreign Nationals by the AAF 1939–1945*. Army Air Forces Historical Studies No. 64. Army Air Forces Headquarters: Air Historical Office. Copy available in the USAF Historical Archive, Maxwell AFB, Alabama.

U.S. Army Command and General Staff College (CGSC *Hall of Fame*). 2011. International Hall of Fame Ceremony, 3 October 2011. Online at: http://usacac.army.mil/cac2/cgsc/events/ihof/2011/index20111006.asp; accessed on 6 February 2012.

U.S. Army Command and General Staff College (CGSC). 2009. *U.S. Army's Command and General Staff College International Military Student Program*. Command briefing, February 2009.

U.S. Army War College. 2010. *Historic Carlisle Barracks*. Online at: www.carlisle.army.mil/usawc/about/Historic.cfm; accessed on 21 September 2010.

U.S. Central Intelligence Agency (CIA). 2012. *The World Fact Book*. Online at: www.cia.gov/library/publications/the-world-factbook/index.html; accessed on 7–9 May 2012.

U.S. Central Intelligence Agency (CIA). 1981. *The Development of Soviet Military Power: Trends since 1965 and Prospects for the 1980s*. Intelligence Assessment, SR 81–10035X, April 1981. From a collection of recently declassified documents from the U.S. Central Intelligence Agency, 2001, CIA's *Analysis of the Soviet Union, 1947–1991*. Available electronically at https://www.cia.gov/library/center-for-the-study-of-intelligence/csi-publications/books-and-monographs/cias-analysis-of-the-soviet-union-1947-1991/index.html. Accessed on 29 April 2008 and 26 August 2011.

U.S. Chairman of the Joint Chiefs of Staff. 1997. *National Military Strategy: Shape, Respond, Prepare Now—A Military Strategy for a New Era*.

U.S. Congress. 2010. *Legislation on Foreign Relations through 2008*. Volume I-A of Current Legislation and Related Executive Orders. Washington, DC: U.S. Government Printing Office, March 2010. Online at: http://WWW.DSCA.MIL/programs.LPA/2010/faa_aeca.pdf; accessed on 7 February 2012.

U.S. Department of Agriculture, Economic Research Service (USDA/ERS). 2012. Data Sets. Online at: www.ers.usda.gov/data/; accessed on numerous occasions.

U.S. Department of the Army, Headquarters. 1976. *Field Manual (FM) No.100-5*. Washington, DC: Headquarters Department of the Army, 1 July 1976. A digitized version can be obtained through the Combined Arms Research Library, U.S. Command and General Staff College, Fort Leavenworth, KS.

U.S. Department of Defense (DoD *Total Military Personnel*). 2013. *Total Military Personnel and Dependent End Strength as of December 31, 2012*. Produced by Defense Manpower Data Center on 8 February 2013. Online at: http://siadapp.dmdc.osd.mil/personnel/MILITARY/miltop.htm; accessed on 30 March 2013.

U.S. Department of Defense (DoD *Personnel Strengths*). 2010. *Active Duty Military Personnel Strengths by Regional Area and by Country (309A) June 30, 2010*. Online at: http://siadapp.dmdc.osd.mil/personnel/ MILITARY/history/hst1006.pdf; accessed on 16 November 2010.

U.S. Department of Defense (DoD *QDR*). 2010. *Quadrennial Defense Review Report*. Washington, DC: Office of the Secretary of Defense. Online at: http://www.defense.gov/qdr/images/QDR_as_of_12Feb10_1000.pdf; accessed on 8 November 2011.

U.S. Department of Defense (DoD). 2006. "United States Field Studies Program (FSP) for International Military and Civilian Students and Military-Sponsored Visitors." Department of Defense Instruction Number 5410.17, 15 September 2006.

U.S. Department of Defense (DoD). 1950. *Deployment of Military Personnel by Country as of 30 June 1950.* Online at: http://siadapp.dmdc.osd.mil/personnel/MILITARY/history/309hist.htm; accessed on 16 November 2010.

U.S. Department of Defense, Office of the Secretary of Defense. 1995. *Annual Report to the President and the Congress.* Washington, DC: US Government Printing Office.

U.S. Department of Defense Security Cooperation Agency. 2008. *Foreign Military Sales, Foreign Military Construction Sales and Other Security Cooperation Historical Facts: As of September 30, 2008.* Online at: http://www.dsca.osd.mil/programs/biz-ops/factsbook/FactsBook08.pdf; accessed on 24 September 2010.

U.S. Department of Defense and U.S. Department of State (DoD/DoS *Foreign Military Training*). 2011. *Foreign Military Training, Fiscal Years 2010 and 2011.* Joint Report to Congress, Volume I. Online at: www.state.gov/t/pm/rls/rpt/fmtrpt/2011/index.htm; accessed on 8 November 2011.

U.S. Department of Defense and U.S. Department of State (DoD/DoS). 2008. *Foreign Military Training and DoD Engagement Activities.* Washington, DC: Bureau of Political-Military Affairs. Online at: http://www.state.gov/t/pm/rls/rpt/fmtrpt/2008/index.htm; accessed on 24 September 2010.

U.S. Department of Defense and U.S. Department of State (DoD/DoS). 2007. *Foreign Military Training: Joint Report to Congress, Fiscal Years 2006 and 2007.* Washington, DC: Bureau of Political-Military Affairs. Online at: www.state.gov/t/pm/rls/rpt/fmtrpt/2007; accessed on 21 January 2008.

U.S. Department of State (DoS *Congressional Budget Justification*). 2010. *Congressional Budget Justification Volume 2 Foreign Operations Fiscal Year 2011.* Online at: http://www.state.gov/documents/organization/137936.pdf; accessed on 29 December 2010.

U.S. Department of State, Bureau of Political-Military Affairs. 2008. *Foreign Military Training and DoD Engagement Activities of Interest, II. Description of Programs.* Online at: www.state.gov/t/pm/rls/rpt/fmtrpt/2008/126353.htm; accessed on 18 November 2010.

U.S. Government. 2004. *The 9-11 Commission Report: Final Report of the National Commission on Terrorist Attacks upon the United States.* Official Government Edition. Washington, DC: US Government Printing Office.

U.S. Government Accountability Office (U.S. GAO). 2011. *International Military Education and Training: Agencies Should Emphasize Human Rights Training and Improve Evaluations.* GAO-12-123 Report to Congressional Committees. Washington, DC: U.S. Government Accountability Office.

U.S. Marine Corps University (USMC). 2010. *Marine Corps University History.* Online at: http://www.mcu.usmc.mil/Presidents%20Document%20Library/MCU%20History2.pdf; accessed on 23 September 2010.

U.S. Naval War College (U.S. Naval WC *Fast Facts*). 2012. *Fast Facts.* Online at: http://www.usnwc.edu/About/Fast-Facts.aspx; accessed on 18 June 2012. These statistics were current as of the last website update in January 2012.

U.S. Naval War College (U.S. Naval WC *History*). 2012. *History of International Programs at NWC.* Online at: http://www.usnwc.edu/Departments---Colleges/International-Programs.aspx; accessed on 18 June 2012.

U.S. Naval War College (U.S. Naval WC *Programs*). 2012. *International Programs at the Naval War College*. Online at: http://www.usnwc.edu/Departments---Colleges/International-Programs.aspx; accessed on 18 June 2012.

U.S. White House. 2010. *National Security Strategy.* Washington, DC: White House.

U.S. White House. 2006. *The National Security Strategy of the United States of America.* Washington, DC: White House.

U.S. White House. 2002. *The National Security Strategy of the United States of America.* Washington, DC: White House.

U.S. White House. 1993. *National Security Strategy of the United States of America.* Washington, DC: White House.

U.S. White House. 1987. *National Security Strategy of the United States.* Washington, DC: White House.

Waheed, Zaklyyah. 1993. "Russian Army Refocuses Its Attention." *Fort Leavenworth Lamp*, 29 January 1993, 15.

Wall Street Journal. 2009. "Not So 'Smart Power': Congress Sticks a Gratuitous Thumb in Pakistan's Eye." *Wall Street Journal*, 15 October 2009.

Waltz, Kenneth N. 1979. *Theory of International Politics.* New York: McGraw Hill.

Weigley, Russell F. 1977. *The American Way of War: A History of United States Military Strategy and Policy*. Bloomington, IN: Indiana University Press.

Wendt, Alexander. 1999. *A Social Theory of International Politics.* New York: Cambridge University Press.

Willard, Robert F. 2010. *Statement of Admiral Robert F. Willard, U.S. Navy, Commander of U.S. Pacific Command, Before the US Senate Armed Services Committee on US Pacific Command Posture, March 24, 2010*, p. 5. Online at: http://www.pacom.mil/web/pacom_resources/pdf/Willard_Statement_SASC_032610.pdf; accessed on 18 August 2010.

Williams, Robert (Major General, Commandant US Army War College). 2010. *Army War College Graduation Address.* Carlisle Barracks, Pennsylvania. Online at: http://www.usgovernment.tv/military/army-war-college/2010/06/2010-army-war-college-graduation/, accessed on 22 September 2010.

Wilson, Woodrow. 1917. *Declaration of War Message to Congress, April 2, 1917.* Records of the United States Senate, Record Group 46, U.S. National Archives. Online at: http://ourdocuments.gov/doc.php?flash=true&doc=61; accessed on 30 November 2011.

Wright, Kimberly L. 2010. "ACSC graduates embrace coalition mindset." *Inside Maxwell AFB*, 15 June 2010. Online at: http://www.maxwell.af.mil/news/story.asp?id=1232-09356; accessed on 24 September 2010.

Wright, Robin. 2004. "U.S. Struggles to Win Hearts, Minds in the Muslim World. Diplomacy Efforts Lack Funds, Follow-Through." *The Washington Post*, 20 August 2004, A01.

Ye, Weili. 2001. *Seeking Modernity in China's Name: Chinese Students in the United States 1900–1927.* Stanford, CA: Stanford University Press.

Yudhoyono, Susilo Bambang. 2005. Speech for the International Hall of Fame Award. Fort Leavenworth, Kansas, 12 September 2005. Online at: http://www.disam.dsca.mil/pubs/Vol%2028_1/Yudhoyono.pdf; accessed on 19 June 2012.

Zimmerman, Thomas. 2011. "Crocker, Martin, Challenge Army War College Grads to Lead in Uncertain Times." Army War College Public Affairs. Online at: http://www.carlisle.army.mil/banner/article.cfm?id=2032; accessed on 6 February 2012.

Index

About the Author

Carol Atkinson is a retired military officer and professor of the practice of international relations at the University of Southern California. In fall 2013 she was a Fulbright Fellow in Sofia, Bulgaria, at the Rakovski National Defence Academy where she continued her research on military exchange programs. Dr. Atkinson holds a PhD in political science from Duke University, an MA in geography from Indiana University, an MA in military operational art and science from the U.S. Air Force's Air Command and Staff College, and a BS in geography/international affairs/Russian area studies from the United States Air Force Academy.

While in the military, she served in a wide range of operational assignments in the fields of intelligence, targeting, and combat assessment. She is a veteran of Operation Desert Storm (1991) where she served on the central command headquarters' staff in Riyadh, and, subsequently, on the contingency planning staff in Dhahran/Khobar, Saudi Arabia. During the Cold War, she was an intelligence advisor on Strategic Air Command's nuclear airborne command post, an imagery analyst at the Foreign Technology Division, and a target analyst in Korea.

Dr. Atkinson's interest in military-to-military exchange programs stems from her experiences working with foreign military officers both abroad and in the United States. Dr. Atkinson teaches courses on U.S. national security, international politics, warfare and emerging threats, and the role of ideas, norms, and identity in international security. She has taught at the U.S. Air Force Academy, U.S. Air Force Air Command and Staff College, University of Southern California, Stanford University, and Vanderbilt University. In her spare time, she enjoys long distance running, hiking, camping, and cross-country skiing.

Lightning Source UK Ltd.
Milton Keynes UK
UKOW03n1419200614

233794UK00001B/24/P

9 781442 231283